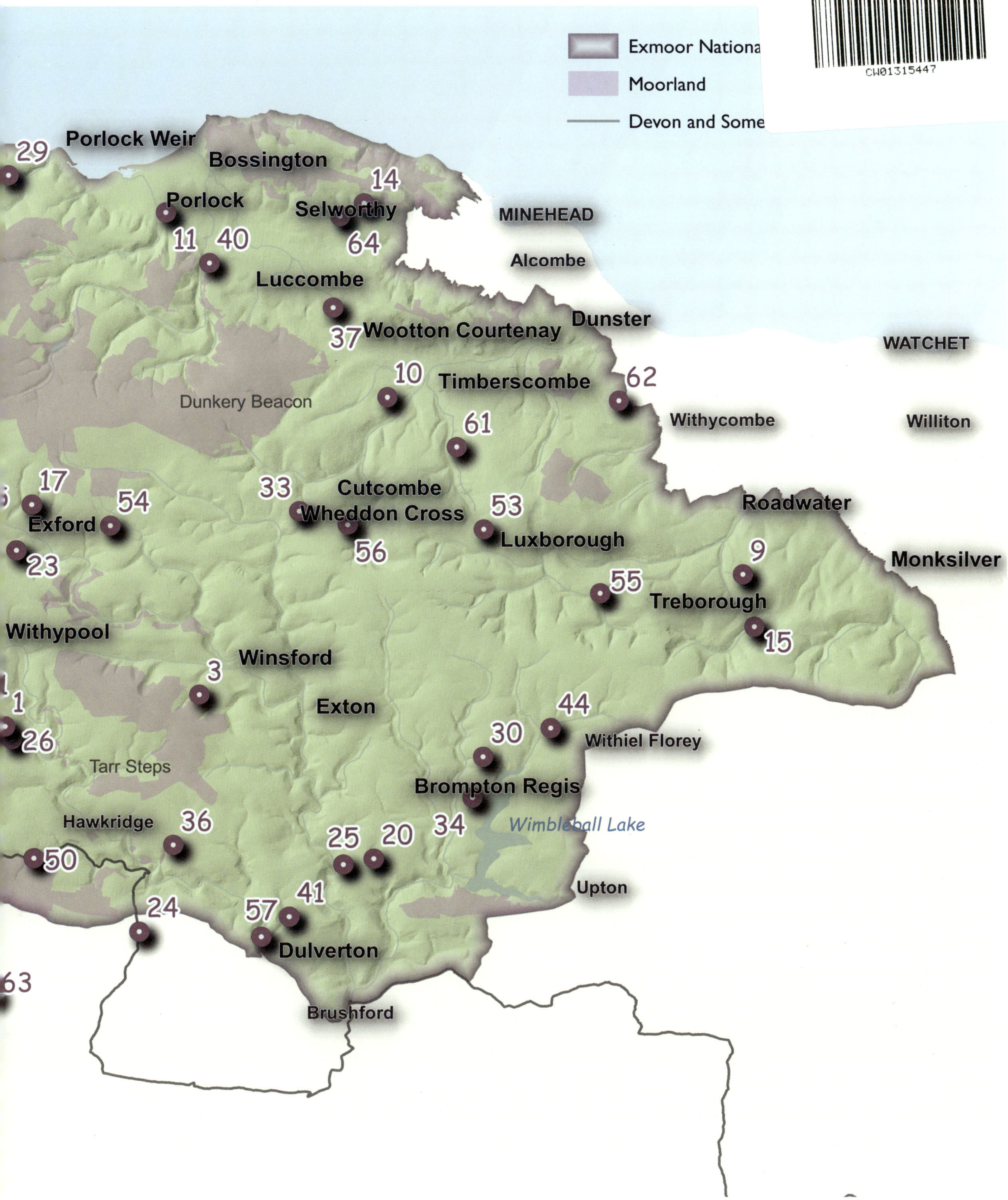
Exmoor Nationa
Moorland
Devon and Some
29
Porlock Weir
Bossington
14
Porlock
Selworthy
MINEHEAD
11
40
64
Alcombe
Luccombe
Dunster
37
Wootton Courtenay
WATCHET
10
Timberscombe
62
Dunkery Beacon
Withycombe
Williton
61
17
33
Cutcombe
54
Wheddon Cross
53
Roadwater
Exford
Luxborough
23
56
9
Monksilver
55
Treborough
Withypool
15
Winsford
3
1
Exton
44
26
30
Withiel Florey
Tarr Steps
Brompton Regis
Hawkridge
36
34
Wimbleball Lake
25
20
50
Upton
24
41
57
Dulverton
63
Brushford

Phil Stocker, Chief Executive, The National Sheep Association

'A picture says a thousand words, and this book, a year long story created with a collection of remarkable and deeply meaningful photographs of Exmoor farming life, couldn't be a truer example. The story is evidence of the passion and "sense of being" of the people who live and work in this tough but stunningly beautiful area, who use the land to carve out a living – and live and breathe responsibility for a special landscape that is treasured and valued by many. We live in a world of dichotomy, where people are interested in how land is managed and food is produced, yet are increasingly disconnected from the practicalities. So the importance of communicating this Exmoor farming world is more important now than ever.

But we should also allow the farming and rural community to indulge for a while in some of the scenes they have created, and take time to celebrate what they have achieved as they continue to cradle traditional values in a world of rapid change. This work marks 70 years of National Park status on Exmoor, and celebrates 10 years of highly valued farmer collaboration through the Exmoor Hill Farming Network, a truly independent group that has helped over 450 farming businesses to adapt and succeed in their business ventures, supporting rural employment and enhancing the landscape. Long may it all continue.'

Baroness Ann Mallalieu, KC

'This book is about Exmoor and its people today. One of our smaller National Parks and, as yet, still largely unchanged by mass tourism, it is one of the relatively few deeply rural communities left in England.

Agriculture is still its underlying foundation. The seasons and the weather dictate the pattern of the life in a way which is no longer the case in our largely urban and suburban country. Exmoor is not a wilderness. A range of different landscapes, all within a relatively small radius, are the work of those who farmed here in the past and those who do so today. They have done it brilliantly. It is so beautiful it makes you cry.

Our world is changing very fast and so is Exmoor. Government policies, changing farming practices, new technology, new ideas and new people make that inevitable. The challenge now is to keep the unique spirit of this working farming community and its incomparable scenery for future generations. Some of those who are rising to this challenge are celebrated in this book. We should all be grateful to them.'

EXMOOR FARMS
A Year on the Moor

by Victoria Eveleigh

with photography by Eleanor Davis

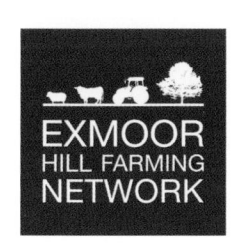

First published in 2024

Copyright © The Exmoor Hill Farming Network/The Exmoor Society and the author and photographers in respect of their specific contributions.

Cover Photograph: View from Anstey Common towards Zeal Farm, Hawkridge

Photographs: All photographs are provided by Eleanor Davis, unless stated otherwise.

Additional photographs supplied by: Exmoor Hill Farming Network (EHFN), Victoria Eveleigh (VE), Chris Eveleigh (CE), Sarah Eveleigh (SE), Emily Fleur Photography, Shaun Davey, Sarah Hailstone, Emily Whitfield–Wicks, Debbie Tucker and Melanie Davies as well as other farmer contributions.

Drawings by Jen Brookes

All rights reserved. No part of this publication may be reproduced, stored in a retrieval system, or transmitted in any form or by any means, electronic, photocopying, recording or otherwise, without the prior permission of the publishers EHFN and ES.

ISBN 978-1-9997330-9-4

British Library Cataloguing-in-Publication-Data
CIP data for this book is available from the British Library.

Sales enquiries to:
Exmoor Hill Farming Network Tel: 01643 841455 Email: admin@ehfn.org.uk
Website: www.ehfn.org.uk

EXMOOR FARMS A Year on the Moor is a partnership between
The Exmoor Hill Farming Network (EHFN) and The Exmoor Society (ES).

We would like to acknowledge our longstanding relationship with

Typeset, printed and bound by Short Run Press, Exeter

Dedication

Robert Deane
1964 - 2023

We are indebted to Robert's dedication to Exmoor and the support he gave to the Exmoor Hill Farming Network, the Exmoor National Park Authority and The Exmoor Society.

Contents

Foreword: p6

April: p10
Worth Farm, Lambscombe Farm, Halse Farm, Great Combeshead Farm, Westermill Farm, Middle Dean Farm

May: p24
HM King Charles III Coronation, Hall Farm, Andy Jerrard, The Woolhanger Farming Partnership, Farm Plastic Recycling, Wood Advent Farm

June: p40
Adam and Oliver Hill, Charlie Rook, Kate South, Exmoor Horn Wool, The Royal Cornwall Show, The South Molton Grading Depot, Wydon Farm, Mill Reef Farm

July: p60
The Exmoor Forest Estate, Sowing the Seeds Project, Meadows Day, Coombe Farm, Ilkerton Ridge Commoners, Moorland Event, Lyncombe Farm

August: p76
Lower Blackland Farm, Bridget Goscomb, The Wellshead Estate, Furzemoor Farm, Chiltons Farm, Oxgrove Farm, Westwater Farm, Devon Closewool Centenary

September: p98
West Whitefield Farm, West Mead Farm, Yarner Farm, Woolcotts Barn, Fran Bullard

October: p114
Exmoor Farmers Livestock Auctions, Lower Woolcotts Farm, Julian Branfield, The Milton Family, The Wallace Family, Woolhanger Sheep Dog Trials, Rabbit Slattery, Harvest Celebrations

Image: Tom Burge's Sheep at Oaremead Farm (February)

November: p134

Jack Croft and Fran Murray, Cranscombe Cleave, Horner Farm, Exmoor Vintage Beef, The Hollam Estate, Oareford Farm

December: p150

Slattenslade Farm, Castle Hill Farm, Shoulsbarrow Farmhouse, Combe Martin Meats, Christmas Fatstock Show, Beech Tree Farm, Anstey YFC Tractor Run, Charmain Dascombe, Singing in the Ring

January: p168

The Molland Estate, Warren Farm, Zeal Farm, Oaremead Farm, East Middleton Farm, James Gregory

February: p182

Patrick Kift, Stetfold Rocks Farm, Geoffrey Illing, Lower Court Farm, Helen and Sarah Thomas, Shearwell Data Ltd, Lucy Gill, The Farming in Protected Landscapes Team, Girt Down Farm, Pre-Lambing Breakfast, South Heasley Farm, The Exmoor and District Deer Management Society

March: p 202

Great Champson Farm, Croydon House Farm, Briddicott Farm, Robin Milton, Hindon Farm, The Reverend Prebendary David Weir

The Team: p218
Acknowledgements: p220
Sponsors and Grants: p221
Glossary: p222
Acronyms : p224

Foreword

Above: L to R: Kate, Sarah and Ian.

When Exmoor was designated a National Park in 1954, it was in recognition of not only the natural beauty of its landscapes but also the cultural heritage shaped by those who have lived and worked here over centuries. People have inhabited Exmoor since prehistoric times, leaving their mark on the moors, hedgerows, woodlands and farmlands that define this area. The biodiversity found here is the result of the interaction between people and nature over thousands of years and the generations of farmers and land managers who have made a living as the land's custodians.

Farming on Exmoor may appear timeless, but it has always been subject to change; agriculture evolves with new opportunities and pressures, shaping how farming businesses develop. Today, the pace of change feels quicker. Farmers must adapt to new regulations, techniques and consumer demands whilst maintaining the environment, combating climate change and ensuring their businesses are sustainable and profitable.

This book, *Exmoor Farms: A Year on the Moor,* offers a glimpse into the lives of those who live and work on the moor. It highlights the great variety of farm types and people, their passions, and their challenges. While most farms focus on livestock production, each farmer has their unique way of making a living. They diversify and branch out – creating, and underpinning, a varied range of ancillary businesses that support and enrich life on Exmoor.

Historically, Exmoor has been home to breeds like the Exmoor Horn sheep, Red Devon cattle and Exmoor ponies and these breeds still thrive – often raised using methods unchanged for centuries. However, farmers are also embracing new technologies and techniques. They use satellite tracking collars on cattle to manage grazing and develop new sheep breeds like the Exlana, which sheds its wool and requires less intervention. While food production remains central to Exmoor's farming businesses, this is done with increasing concern for the wider environment. Farming methods are developing to enrich biodiversity and reinforce Exmoor's populations of rare and iconic species – from red deer and grey long-eared bats to nationally significant populations of lichens and fungi.

Exmoor life is enriched in other ways too. Farming can be isolating but there is a strong, close-knit community on the moor – a community ranging from neighbours to the broader networks and organisations that cooperate to strengthen farming life. One of these is the Exmoor Hill Farming Network, established in 2014 to support upland farmers and improve the viability of their businesses through knowledge-exchange, peer-group meetings and training. The Exmoor Society, formed in 1958, works to conserve the National Park by promoting sustainable farming and understanding of the Exmoor environment. And Exmoor National Park Authority works closely with the farming community to carry out the twin purposes of our national parks: conservation and access. All three are driven by a shared desire to champion the Exmoor farming community.

Celebrating the 70th anniversary of Exmoor National Park's creation, and the 10th anniversary of the Exmoor Hill Farming Network, *Exmoor Farms* is not just a collection of portraits; it is a record of the enduring relationship between humans and the land seen through the cycle of a farming year. It captures Exmoor's farming heritage, the resilience of its farmers, and the evolving landscape of agriculture in this region. Turning the pages you understand the deep sense of place that ties people and land together and witness the efforts taken to protect Exmoor for future generations.

Exmoor Farms: A Year on the Moor is a tribute to the farmers of Exmoor, whose stories are as varied as the land itself, ensuring that this remarkable landscape continues to thrive.

Kate O'Sullivan
Chair,
The Exmoor Society

Sarah Bryan
Chief Executive,
Exmoor National Park Authority

Ian May
Chairman,
Exmoor Hill Farming Network

September 2024

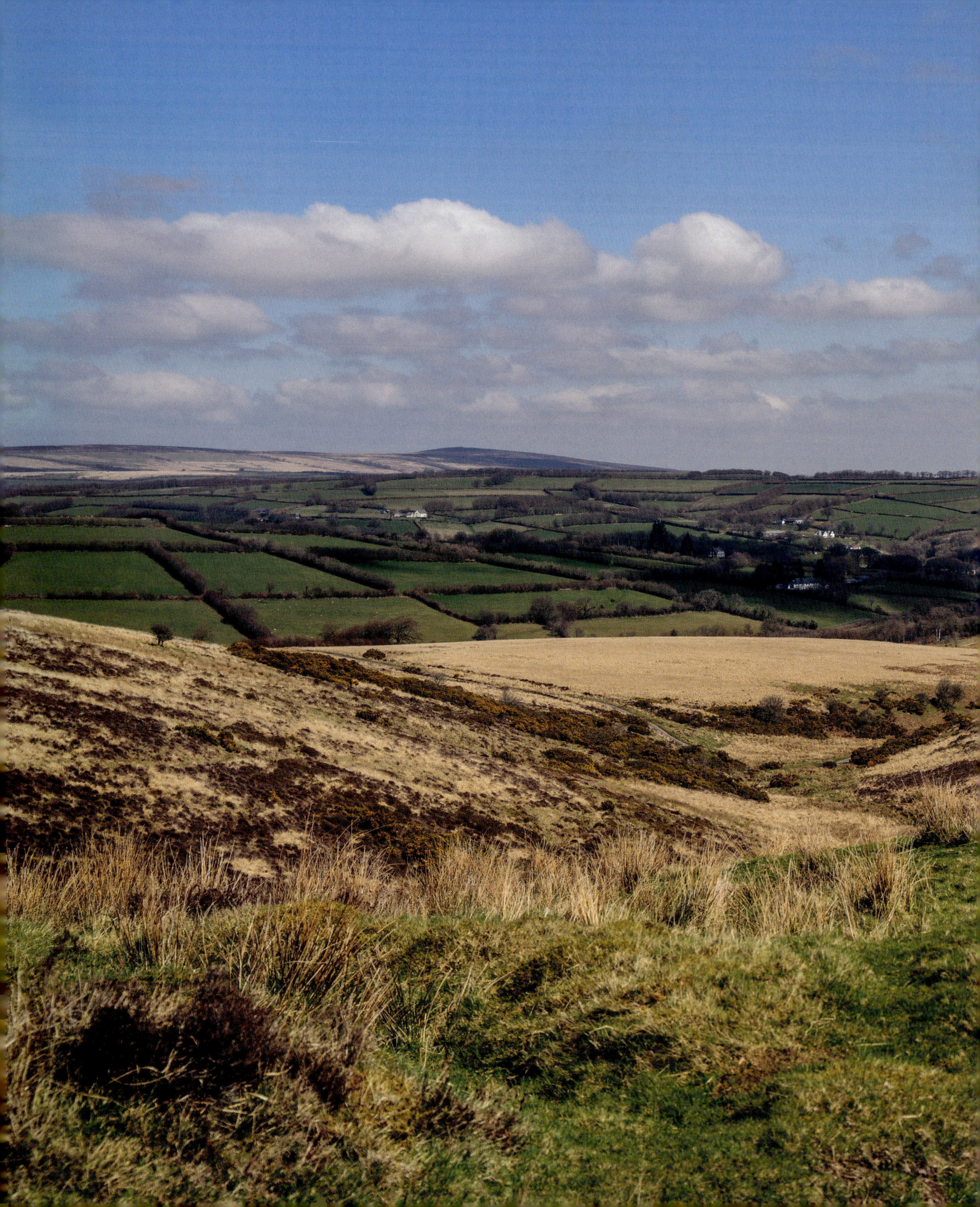

View from Withypool Common.

April

Spring always comes late to the hills of Exmoor. However, this year it was particularly cold and wet. April means lambing for many. Sheep typically give birth in the spring, so it makes sense to fit in with this natural cycle.

Lambing can be exhausting and stressful, especially when the weather is bad. On some farms – like Worth, Lambscombe and Halse – ewes give birth in a shed and are let out with their lambs within the next few days, while on other farms lambing takes place outside. At Westermill, outside lambing begins in early April, whereas the Geen family lamb their wool-shedding Exlana ewes a couple of weeks later.

Most farmers on Exmoor breed cattle as well as sheep, but modern cattle can calve all year round so it is easier for farmers to choose when calving will take place. On some farms it coincides with lambing, but on others it happens at another time to make full use of available forage, shed space and labour, or to supply a specific type of market.

Calving will feature in other months, but for now the spotlight is on lambing.

Alan Collins and Ann May
Worth Farm, Withypool

At 85 years old, Alan Collins has been through more lambing seasons than most farmers on Exmoor. "It's not a chore – I look forward to it," he says. "It's important to have a good team. This year there's my daughter Ann, who lives here during lambing; Robert, who works on the farm full time; Daniel and Molly, who are day people; Harry, a local young man who does nights; and we also have a cook for about four weeks. I do the paperwork, feed the cattle and sheep and act as the gofer, driving to the vets or Mole Valley Farmers for supplies."

He adds that it's easy to justify extra help when there are over 1,000 ewes to lamb, whereas farmers with small flocks tend to work day and night, and end up exhausted.

Alan and Ann do their lambing indoors from the last week in March to around the third week in April at Worth, and the lambs are taken out to the fields with their mothers when they're strong enough. The attention to detail is second to none.

Alan farms about 1,000 acres at Worth Farm, Withypool, and Hill Farm, Hawkridge, rents around 100 acres at Brightworthy Farm and has grazing rights on Withypool Common. However, he wasn't born into farming. His parents sold government surplus clothing in Taunton after the Second World War, but he wanted to work in the countryside rather than being a shopkeeper. So – after two years of National Service with the King's Troop Royal Horse Artillery, where he "learned to ride properly" – he started his farming career with 10 Saddleback sows and a boar in some fenced-off woodland at Southill Farm, Withypool, which was owned by his uncle.

Before long, Alan was renting Knighton Farm from his uncle. He became friendly with Dave and Fred Rawle, who owned and trained horses, and started riding in point-to-points. Soon he'd bought a horse of his own from a sale in Exeter.

"Topper was cheap because he wouldn't jump, but I'm very interested in horse psychology, and I enjoy a challenge. I took things right back to the beginning, to build

Below: Alan and Ann, father and daughter, together in the lambing shed.

up his confidence, and we started winning races together. We won two Maidens in a week." Alan's face lights up with the memory of it. "And a few weeks later I had a fantastic race: Bertie Hill and I were neck and neck over the last fence, going like the clappers, and I won!"

Friends used to tease Alan that he'd "got the nest but not the bird", but it wasn't long before he met May Westcott at Young Farmers. They married in 1961, and their daughter Ann was born a couple of years later. By that time, they owned Knighton.

Alan carried on riding in point-to-points until May was expecting their daughter Ann. "The two of us had a committee meeting, and May said I'd best concentrate on the farming." With typical good humour, he adds, "I had a couple of lucky years and quit while I was ahead."

One day, in 1970, the postman brought news that Worth Farm was for sale. Alan and May had another committee meeting and decided to be bold. They completed the sale of Knighton and bought Worth on the same day.

"Land was a hundred pounds an acre and bank loans were easy to negotiate," Alan recalls. "Fairly self-sufficient farms of around fifty acres were common, but times were changing and smaller farms were being sold, so we took the opportunity to buy land as it came up… I'm afraid Exmoor will lose more farms in the next few years, with the changes going on. We knew where we were and what we were going to get with the Basic Payment Scheme. It's always been a job to plan in farming, but the uncertainty around Environmental Land Management [ELM] is making things much worse."

When The Rt Hon Mark Spencer MP, Minister of State for Food, Farming and Fisheries, visited Worth Farm with Katherine Williams and Ian May from the Exmoor Hill Farming Network (EHFN) at the beginning of April, they put their concerns to him. "We told him you've got to have a fair-sized farm these days to survive, and he knew what we were talking about," Alan says.

He is no stranger to government at a local level. He has been a Withypool and Hawkridge Parish Councillor since 1964, Clerk for 10 years and Chair for 24. He was a Rural District Councillor for Dulverton for several years, too, and Chair of the Highways Committee. He has also been Chair of the Village Hall Committee for several years. In the mid-1990s, he became a Parish Council Member of the Exmoor National Park Authority (ENPA) and ended up being Chair of the Finance and Resources Committee.

"We started off with a meeting a month and ended up with two meetings a week. I had to get up early to do the farm work beforehand, but it was very interesting. I learned a lot and saw Exmoor from a different perspective."

May Collins died in 2020, and the farm is now "under new

Below: Inside the lambing shed, with expectant ewes in the foreground, ewes and their newborn lambs in individual pens and, to the right of the picture, ewes and older lambs waiting to go outside.

From top left clockwise: An Exmoor Horn ewe in the early stages of giving birth; Molly Matthews using a stomach tube to feed colostrum to a lamb that's too weak to suck; a Scotch Blackface ewe licking her newborn lamb clean; Ann bottle-feeding a lamb under the watchful eye of its mother.

management", as Alan puts it, because Ann has become a partner. She also farms near Barnstaple with her husband and daughter, so she's incredibly busy.

They haven't changed things much, but one thing Ann wanted to do was to sell the Scotch Blackface ewes that had been grazing Withypool Common and replace them with Exmoor Horns.

"Ann likes Exmoor Horns, and the lambs sell better," Alan explains. "We bought quite a few last autumn. I enjoy going to market with her because we like the same things. She does the bidding, but sometimes she'll look at me if the price is rising, and I'll say, 'Go on! Go on! Take it home!' You've got to be brave if there's something you like." He gives a wry smile. "Well, you've got to have the money, actually. Let's put it this way: when you're buying breeding stock, buy the best you can afford."

Some of Alan and Ann's Exmoor Horns are used to produce pure-bred lambs and the rest are crossed with Bluefaced Leicester rams. Exmoor Mule ewe lambs from Worth are sought-after at the autumn sales. Several Bluefaced Leicester rams are also bred every year at Worth, with the aim of producing a hardier, broader-backed type that can cope with Exmoor's weather.

Over the years, Exmoor Horns have been bred for cleaner faces and legs and a larger frame. Their feisty characters haven't changed much, though. Alan was reminded of that when he broke several ribs trying to catch an Exmoor ewe at lambing time in 2022. It's thought a blood clot from that injury gave him a stroke a few weeks later.

"Luckily, Robert was on the quad bike with me when it happened," Alan says. "He drove me into Barnstaple Hospital, and I had six days' full board with nurses looking after me. I really enjoyed myself! It was the most I'd ever been on holiday – before that I'd only had a couple of days up north buying sheep."

Calving comes straight after lambing at Worth, and it takes place out in the fields. They ran 100 Angus-cross-Friesian cows with a Charolais bull last year, and aim to sell the calves as newly weaned suckled calves in the Native Breeds Sale at Cutcombe Market in November. Bovine tuberculosis (usually referred to as TB) is always a worry, as cattle must have a clear test to be accepted for sale. "Luckily, we've been clear of TB for twelve years now. It may be something to do with this area being one of the first to have a badger cull," Alan says. "There are lots of deer around at the moment, though, so I hope they haven't got TB."

Like most farmers, Alan has had a lifelong interest in the wildlife on his land: "It's in the back of my mind with everything I do."

He particularly looks forward to the swallows returning, and to hearing the first cuckoo. One cuckoo uses a nest in the same gorse bush every year, and a family of kestrels has settled at the farm. "They're lovely to see, but they've killed the sparrows. That's the thing about nature: what's good for one thing may not be so good for another. As with most things, it's all a question of balance."

Worth Farm has grazing rights on Withypool Common, and grazes 45 cows from 1st May each year and 100 ewes and lambs from 14th May, then 287 ewes after weaning. The control of gorse and purple moor grass (also known as *Molinia*) has been a major issue.

"Gorse is like a cancer: you don't notice it until it's bad," Alan observes. "We've found you get the best results by cutting it first with a flail, which smashes the stumps so that you don't need to use any chemicals, and then mowing any shorter plants regularly so they don't get the chance to flower. There are acres and acres of *Molinia* as well. Burning followed by grazing helps to control that. Heather is gradually coming back, but the heather beetle is a problem."

Alan has seen some remarkable changes in farming since he started in the late 1950s – not so much in essential hands-on stockmanship skills but in the scale of farming, the demise of small mixed farms, government policies, the amount of record-keeping involved and, most of all, technology. "When I started farming, we used horses for stock work, then it was a Fergie tractor followed by larger tractors – which made a mess – and now quad bikes. They're convenient, quick, kinder to the land and kinder to the dogs because they can hitch a lift until they're needed," he says. "Yes, if I could name one change that's made all the difference, it would be the quad bike."

Below: Rob Fry taking ewes and lambs from the shed to the field. The lambs are in a separate compartment at the front of the trailer to keep them safe.

Below: Alan on his quad bike, checking ewes and lambs in the field.

Below: Hedges shelter ewes and lambs from bad weather and provide a stock-proof field boundary.

Ian May
Lambscombe Farm, North Molton

With just under 500 ewes, Ian, like many farmers on Exmoor, relies on family and friends to help out at lambing time rather than employing anyone. His husband, Ian Fisher, had never done any farming before he moved to Exmoor 10 years ago, but now he's an experienced and essential part of the team.

"I do from eight in the morning until around midnight on a good day, or later if there's lots on," Ian May says, "and then Ian gets up at around three in the morning and is in charge until breakfast time. He's a good cook, so he usually prepares the evening meal, which gives him a bit of a break from lambing in the afternoon. My aunt Ann, who was brought up here at Lambscombe Farm and knows it well, comes most days and takes the ewes and lambs out to the fields for us. And Irene (who used to work with Ian and has remained such a good friend that she volunteers to come lambing every year) stays with us for two to three weeks and helps keep the sheds in order – lambing ewes when needed, feeding lambs, filling water buckets and mucking out lambing pens. My niece Amie loves helping when she can, as does Harriet, the granddaughter of friends of ours. Dad doesn't do lambing anymore, but he provides vital supplies, including flapjacks for us and sheep feed for the ewes."

Above: North Country Mule ewes.

Together they all make a good crew, with Ian at the helm.

"It's interesting that we'd never really thought about our different roles until I had to go to Worth Farm with Katherine to meet Mark Spencer," he says. "When I got home, everything was fine, but my husband said he hadn't realised what a responsibility it was being in charge."

Farming is often seen as a manual job, and in many ways it still is, but running a successful livestock business also involves a huge amount of decision-making, along with other managerial and office skills.

Daily snap decisions are usually based on experience, but planning for the future requires as much information as possible.

Ian finds that keeping detailed farm records helps tremendously.

"I've been taking part in tests and trials for the new Environmental Land Management schemes, the Sustainable Farming Incentive [SFI] pilot and the Health and Welfare Pathway," he says. "I really enjoy engaging in the process and working out what's achievable or impractical."

Left: Ian Fisher helping a ewe to give birth.

In addition, Ian is Chair of the Exmoor Hill Farming Network and Regional Manager of the National Sheep Association.

The ewes at Lambscombe are a mixture of North Country Mules with home-bred Suffolk crosses and Texel crosses: large, prolific sheep that tend to have more lambs but can be less hardy than Exmoor Horns.

Ian likes to get the bulk of his lambing done before the suckler cows start calving in the second week of April, so lambing is from mid-March until mid-April. Usually that isn't a problem, but this year a spell of wet, windy weather came at the worst possible time. Shed space is limited, so ewes and lambs are usually turned out after a couple of days to free up pens for new arrivals.

"I'd rather not dwell on it too much. It's pretty heart-breaking, though, to see good, strong lambs struggling outside in conditions like that. You do everything you can but it's not always enough," Ian says. "This time it was especially tough, but it's always a relief when lambing's over for the year. In fact, there's an adage that if you had to put the rams back in with the ewes straight after lambing then nobody would do it! However, the seasons change quickly and it's important to look forward, address where we can improve things next time and focus on the positives."

Above: If a ewe can't rear all her lambs, one or more are taken away and fed using a hand-held bottle or a lamb feeding system.

Above: Ian May drenching ewes with wormer before they go out to a field with their lambs.

Below: Teamwork. L-R: Harry the dog, Irene, Ann, Ian May, Adam and Amie.

Jeremy and Tish Brown
Halse Farm, Winsford

"I have been known to say that I hate lambing with a passion, but farming's a funny thing – you love it, and you hate it," Jeremy Brown says. "I must admit that this year I really did hate it when Storm Noa came along on the twelfth of April. It was far worse than I thought it would be, and it caught me out as one of the fields we were using didn't have much shelter. We lost eight lambs and had to bring twenty-seven ewes back to the shed because their lambs were so cold."

Jeremy and his wife Tish have Halse Farm, near Winsford, and lamb 380 sheep from late March into April. Like Ian May, they have North Country Mules and home-bred Suffolk cross Mules – large, productive sheep that aren't particularly hardy.

The ewes are housed from early January onwards, to protect them from the weather and give the fields a break. They are also scanned in January, to see how many lambs they're going to have, and this year they scanned at 190% (nearly two lambs to every ewe). Like many farmers, the Browns scan their ewes so that they can feed them according to how many lambs they're carrying, and so they can foster a surplus lamb onto any ewe expecting a single. Usually, there are roughly the same number of triplets as single lambs, so the aim is for each ewe to rear two.

Jeremy and Tish do most of the lambing work themselves, as well as looking after a herd of cattle and running a campsite. Calving starts a few weeks before lambing and sometimes carries on into May, and the campsite is

Below: Jeremy and Tish letting ewes and lambs out into a field of fresh grass.

Above: Jeremy mothering up ewes and lambs after letting them out.

especially busy over the Easter holidays, so there's a lot to do. However, for the main lambing period they're helped by one or two students from Bristol Veterinary School, who have to get practical experience on farms as part of their training.

"We've been lucky with our vet students since we started taking them 28 years ago, and this time we had two very good girls," Jeremy says. "Tish is a morning person and I'm better in the evening, so she gets up at four in the morning and goes to bed after supper and I get up at about eight in the morning and work until about two the following morning. This year, though, the vet students did every other night once they knew what to do, which helped enormously. It's so important for everyone to get some sleep... Actually, what I *really* hate about lambing is the lack of sleep."

Below: Lamb macs can be lifesavers in wet weather.

Below: A couple of lambs enjoying a brief spell of sunshine.

The Geen Family
Great Combeshead and Barton Pitts, Heasley Mill

Bill Geen is one of the farmers who has pioneered a new type of wool-shedding sheep called the Exlana.

He farms in partnership with his son Matt and daughter-in-law Jenny at two farms near Heasley Mill. Bill and his wife Helen live at Great Combeshead – a 260-acre farm rented by his family for over 100 years – while Matt lives with his young family at Barton Pitts, which has 40 acres. They also have 180 acres on Fyldon Ridge and take around 100 acres of seasonal grazing.

Matt and Jenny have a separate business breeding pedigree Hampshire sheep and running a log cabin holiday business at their farm.

Fyldon Ridge is extensively grazed under a Higher Level Stewardship (HLS) agreement, but other, more fertile land is used for rotational grazing, with herbal leys and approximately 100 acres of temporary grass. Around six acres of fodder beet, 20 acres of rape and turnips plus a field each of winter and spring barley are grown to feed the livestock.

The Geens have about 700 Exlana ewes, which lamb outside from mid-April. "It's warmer by then, and the grass is like rocket fuel," Matt says.

They also keep a herd of Stabiliser cattle, which are calved from late spring. Avoiding procedures like dehorning, castrating and tail docking will become increasingly

Below: An Exlana ewe with her twin lambs. The wool on the fence indicates the ewes are already beginning to shed their fleeces.

Above: Bill and Matt Geen.

important in future, they think. Their cattle don't have to be dehorned because they're polled, and neither the calves nor lambs are castrated. Furthermore, there's no need to dock the tails of Exlanas because wool-shedding sheep are unlikely to get flystrike.

"Nearly twenty years ago, I was invited to a meeting with other farmers who were interested in sheep genetics, and at the end of the day about ten of us had formed a group," Bill remembers. "We were primarily interested in producing wool-shedding sheep (hence the name Exlana, as *lana* is Latin for wool), but we've also introduced other

Below: Taking part in an experiment to measure on-farm methane emissions from sheep using portable accumulation chambers. (*The Geen family*)

Above: The Geens' cattle on Fyldon Ridge, looking south towards Dartmoor. *(The Geen family)*

qualities like resistance to worms, good maternal traits and the ability to thrive on a forage-based diet." They formed a company called Sheep Improved Genetics (also known as SIG) and started crossing the best of their own ewes with wool-shedders.

"Peter Delbridge from Blindwell Farm crossed his North Country Mules, for instance, while I crossed my Lleyns," says Bill. "We're a genuine cooperative – a breeding company rather than breed society. It's a long-term relationship based on trust."

"There's lots of recording involved, but it gives us confidence working with people who want the same thing," Matt adds.

The Exlana is an 'open composite breed', with beneficial genetics from all over the world. The sheep in the breeding programme are managed as one flock, although the eight Directors involved have separate farm businesses. One of the Directors is Richard Webber, and SIG has made full use of the electronic identification and data-management systems developed by his company, Shearwell.

"We're now complicating our lives with methane recording, although I suspect the sheep on Exmoor produce a lot less methane than the peat," Bill comments. "SIG is involved in an industry-wide project called CH4nge, measuring on-farm methane emissions from sheep. It's done under licence from the Home Office, so there's lots of form-filling. They bring 'portable accumulation chambers' here for the day, and sheep are put into them for twenty minutes at a time to measure methane emissions. Other measures of production efficiency are noted, and there's subsidised genetic testing because we're trying to identify sheep genetics that can reduce the carbon footprint of sheep farming."

"Everything's changing – we're all moving into unknown territory, with new global challenges and new agricultural support systems," says Matt. "We're already doing things like rotational grazing, herbal leys and minimum tillage, so we'll probably just carry on and try to adapt to whatever the market does. You've got to be adaptable to survive in agriculture."

In June 2023, Bill was elected as a Parish Council Member of the Exmoor National Park Authority.

The Edwards Family
Westermill Farm, Exford

The Edwards family hosted an Exmoor Society walk at Westermill Farm on 17th April, just as their lambing and calving season was drawing to a close. Jill is a Trustee of the Exmoor Society and Oliver is a Director of the EHFN. Their son David helps to run the 600-acre farm and manages the sheep enterprise. The farm is under an HLS scheme that has just been renewed for another five years. It has a well-established campsite and several holiday cottages.

A wide range of people went on the walk, from local farmers to a young couple from London who were on holiday. As everyone went along the valley to the campsite facilities, David met up with them and talked about his involvement with the farm while his sheepdog sat patiently on the quad bike.

Like many young farmers, David spent some time working on farms in New Zealand and came back with ideas based on his experiences there. Lambing at Westermill used to take place indoors, but David has changed to outdoor lambing with a flock of about 500 Perendale sheep, with just a few Cheviots and Mules. Perendales were developed in New Zealand from Cheviots crossed with Romneys.

Below: David and Jill Edwards giving an introductory talk before the walk. *(VE)*

"They're designed for hill country – hardier than Romneys and more prolific than Cheviots," David says. "My goal is to have eight hundred of them."

Lambing outside is cheaper and requires less labour, and if the grass is good enough you don't need to feed concentrates, but the downside is that the flock is vulnerable to bad weather. Predators, especially corvids, can be a problem, too. All in all, you generally lose more sheep if you lamb outside, even if the weather is okay, but David thinks it still makes sense.

The lambs are weaned in August and are all sold as stores to one buyer in September.

Jill led everyone up the valley to a good viewpoint, talking along the way about their campsite and woodland, hedging and fencing, and the loss of trees to ash dieback. She also explained how Oliver's father John was instrumental in starting the ENPA's Moorland Management Agreements in the late 1970s, which paved the way for nationwide agri-environment schemes.

By the time the group was walking past the lambing field and back down to the farm, everyone had got to know each other, and the conversation was wide-ranging; it even included ideas about how the couple from London should clean their previously white trainers.

Below: Walking and talking. *(VE)*

There were plenty of cattle to see in the farm buildings: yearlings and two-year-olds in an outside yard, followed by cows and calves (plus a few sheep needing extra care) in a large cattle shed.

Oliver was there, ready to talk about his herd of 60 pedigree Aberdeen Angus cows, which he has bred for the past 40 years. "I now have cows that suit Westermill," he said, "with a leg at each corner, four good teats that stay the size of a finger throughout their lives, good hips for easy calving and, most important of all, docile temperaments."

He admitted that although he loves farming, especially his cows, at times he finds the work and worry too much, especially when things go wrong – like they did earlier in April when some of the calves got an infection.

When asked why he carries on despite the problems, he replied, "Farming is in my blood, my DNA, on both sides of the family."

Above: Oliver with some of his Angus heifers.

"It's that passion – almost obsessiveness – with the land and livestock that can't be taught in college or from a book," Jill added. "It's a part of who you are."

Above: Aberdeen Angus youngstock watching the walkers go by. *(VE)*

Ginny Kingshotte
Middle Dean Farm, Trentishoe

Monday 24th April was a bleak day. As dark clouds rolled in from the north and icy rain began to fall, farmers clad in waterproofs gathered at Blackmoor Gate for the annual sale of cull ewes and couples run by Exmoor Farmers Livestock Auctions (EFLA).

One farmer and her sheep stood out from the crowd. Ginny Kingshotte, wearing an orange high-viz waterproof jacket, valiantly kept her emotions in check while she talked to prospective buyers about her pens of top-quality Exmoor Horn ewes and lambs – just about coping with factual questions about lambing dates and vaccinations, but fending off fatal expressions of sympathy.

Ginny moved to Middle Dean with her family 15 years ago. To begin with, she and her husband Dave had 200 acres and rented another 40, and they farmed Longhorn cattle and a large flock of Exmoor Horn ewes. In fact, Ginny won the Exmoor Horn Sheep Breeders' Association competition for the best large flock about 10 years ago. Since then, she has also bred and shown prize-winning Exmoor Mules.

Dave suffered acute kidney failure, and developed other serious health issues, a few years after they moved in, so they changed their business by selling the cattle, most of the sheep and all but 50 acres of land. They moved out of the large farmhouse, which they converted to a holiday let, and into a converted dairy next door.

Dave died four years ago.

Planning consent from the ENPA came with various conditions that have complicated implications for both properties. Because of this, plus uncertainty about what the future holds for agriculture and the fact that farming by herself is becoming increasingly difficult due to arthritis, Ginny has reluctantly decided to put Middle Dean on the market.

"It broke my heart to sell my sheep. I've known them all from birth, and last year they did so well, winning cups at Dunster, Exford and Porlock," she said a few days after the sale. "But I'm glad they've gone to people who will look after them and appreciate them."

Below: Ginny's sheep being sold at Blackmoor Gate Market. (VE)

May

At the beginning of the month, the nation's attention turned to London for the Coronation of King Charles III and his wife Queen Camilla. Farmers on Exmoor celebrated this with a get-together at Cutcombe Market.

April's unsettled weather showed little sign of change until the middle of May, when high pressure brought a welcome settled spell with plenty of sunshine.

On several Exmoor farms, including Hall Farm and Woolhanger, calving is now timed to take place outside, in late spring, when it's warmer and the grass is growing. Some of the cattle at Woolhanger graze open moorland, controlled with the help of modern technology.

May is also a good time for an agricultural spring clean; clearing out sheds that have housed animals during the winter and sorting out rubbish like plastic silage wrap so that as much as possible can be recycled. At Wood Advent Farm, they are trying to keep their plastic use to a minimum, but working towards being a zero-waste farm isn't easy.

HM King Charles III
Coronation Celebrations

The day before this historic event, over 60 members of the farming community attended the Farming Equipment and Technology Fund Demonstration afternoon on Friday 5th May. Several local agricultural suppliers were there, including Bridgmans, Harpers Feeds and Masons Kings.

Halfway through, there was a pause in proceedings while everyone celebrated the Coronation. The Reverend Prebendary David Weir cut a magnificent cake made by Edna Hayes, glasses were charged with Pimm's and there was an enthusiastic toast to King Charles III.

As The Prince of Wales, Charles created The Prince's Countryside Fund, now The Royal Countryside Fund. This has helped rural communities in various ways, such as the creation of more than 50 farm support groups through the Farm Support Group Initiative.

"King Charles' support for farmers here on Exmoor has enabled the Hill Farming Network to grow so that we are now helping around 400 farming businesses within the Greater Exmoor area," Katherine Williams said. "We bring people together to share experiences and information while forging friendships. It's so important to tackle rural isolation."

Above: Katherine in the ring at Cutcombe Market, welcoming Network members to the event. *(VE)*

Katherine felt very privileged to attend the Coronation celebrations in London the following day when, with representatives from other farm support networks, she was invited to have a grandstand view of the processions outside Buckingham Palace. "The rain did not dampen our spirits," she said afterwards. "There was a fantastic atmosphere, and it was an experience I'll never forget."

Three days later, Katherine took nine members of the Women in Farming Group on a study trip to Northern Ireland, so she had a busy start to May.

Left: The celebratory cake, made by Edna. *(Edna Hayes)*

UK to Oz, so embryos are the only way to go. So far, the centre has 43 embryos, which the farmer will use to impregnate his cattle in Australia. The other two cows were from my lines, but he especially liked Honour's connection with the King. I've kept her daughter, so I've still got that bloodline, and my herd can return to a high health status."

The cattle are hardy enough to live outside all year round. They have field shelters to protect them from the weather but tend to use them more in summer when it's hot. Karen aims to calve her cows during April and May. The male calves are kept as bulls – some for breeding and some to sell as stores – and the females are either sold for breeding or kept in the herd.

"I find that the cows with longer faces are better milkers whereas the shorter-faced ones are beefier, so I look for something in between," Karen says. "And they must have good udders and feet; I won't have anything with crossed claws or shallow feet."

Her cattle are going from strength to strength. She's got about 30 in total, and has just bought 16 additional acres to keep them on.

Karen doesn't take her livestock to shows, but her open

Below: Karen standing on a bridge she made.

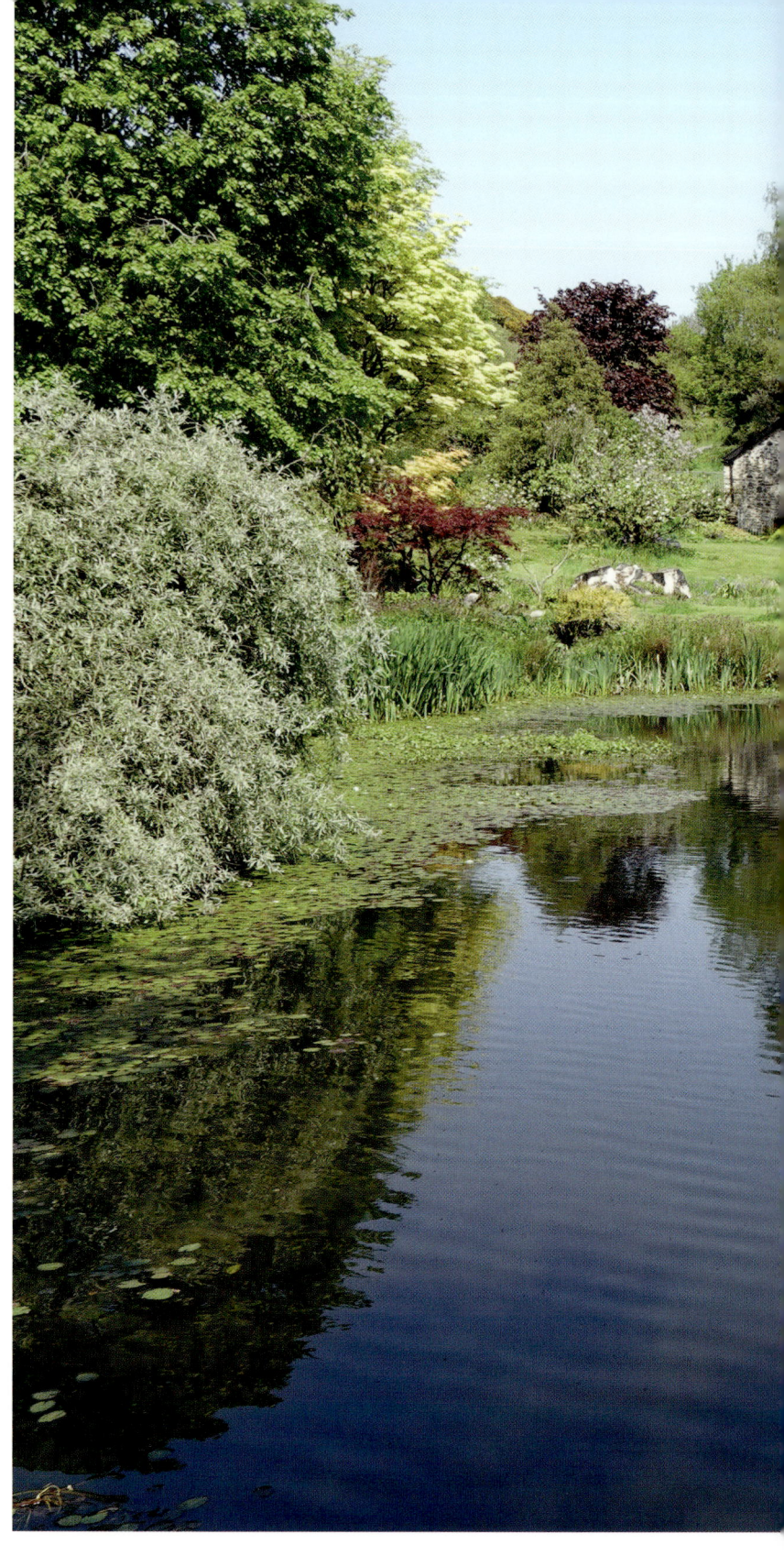

days always generate a lot of interest. There are around 650 breeding female Whitebred Shorthorns worldwide now, and she's glad to have played a part in their recovery.

Looking after the animals was always Karen's job when Nick was alive, but now she has to do the things he did as well, like maintenance work and chopping wood. "Nick used to have a great big Husqvarna chainsaw with a pull cord. I've never been able to master pull cords, so now I've got an electric chainsaw as well as other electric tools. I'm not afraid to give things a go; it's surprising how much of his building knowledge has rubbed off on me," she says.

Help is there when she needs it, though. "Last April, I was in hospital for ten days with shingles, which turned into encephalitis. I was so ill I didn't realise what was going on, but friends and neighbours looked after everything for me without being asked. It just shows what community spirit can do." Karen smiles as she looks at her grandchildren playing in the garden. "I really love it here. I wouldn't want to live anywhere else."

Above: Hall Farm. *(VE)*

Andy Jerrard
Market Chaplain

"The people in charge of agricultural policies at the moment don't seem to realise how deep the emotions are of people who farm the land," Andy Jerrard says. "If you've been brought up to produce food, to undo all the work that generations have done before you can cause huge emotional trauma."

He remembers working with his grandfather, George Huxtable, at Hall Farm, Brendon, and Town Farm, Countisbury. "I particularly remember walling in Century Field at Town Farm with my grandfather and his workers. I gather the National Trust has got plans to let a lot of their hedgebanks go to create wildlife corridors, but I'll be very upset if that wall goes. It's a small thing compared with what a lot of farmers are having to contend with, but it means a lot to me."

Andy is especially worried about how some farmers are being forced to give up their tenancies, both locally and nationally, because their landlords want to go down the rewilding route. His concerns were heard by the District Synod and then went to the Methodist National Conference. "A letter was written to George Eustace, the subject had a fair airing and I'm now an advocate for rural issues as well as Market Chaplain. It's certainly not an easy role because sometimes I get quite cross!" says mild-mannered Andy.

Surprisingly, he had no particular qualifications for the job he does now, other than an interest in people and an agricultural background: school holidays spent at Hall Farm, an HND in Agriculture from Seale Hayne, four years as a herdsman on a dairy farm and then, after he was married, a couple of jobs as a seeds salesman before joining the Environment Agency's monitoring department.

By chance, Andy stopped at a farm shop for a cup of tea one day and saw an advertisement in a magazine for a rural support worker.

"Alan, my predecessor, started just before the foot and mouth crisis. He helped many people through that period and generated a lot of goodwill," says Andy. "My job initially involved pastoral visiting, but there's a limit to how many people you can reach with that, so we thought about how I could be available to anyone in need on a regular basis, and the answer was to go to the markets."

Markets are vitally important places for farmers to meet, especially now that mechanisation is making farming such a lonely job. Andy covers 11 markets in Devon, West Somerset and part of Cornwall. He also attends other events if possible – like the Coronation Party at Cutcombe. The post is funded primarily by donations and grants from within the Methodist Church, with additional contributions from individuals and the agricultural industry. However, he is keen to emphasise that his job is to listen and help where he can, and not to talk about religion or give marketing advice!

Andy believes we are all unique; similar circumstances affect people in very different ways, so nobody can say they know how somebody else feels. "It's all about letting

Left: "Some people don't grasp the pull a piece of land can have on you," Andy says. *(VE)*

people tell their story. Talking often allows them to see things from a different perspective. I try to do my best for people and get them additional help if it's needed. It's the sort of thing churches used to do everywhere. In fact, if every vicar and minister were able to be a David Weir, I'd probably be out of a job… And I definitely couldn't do what I do without Cathy, my wife."

After going to Blackmoor Gate Market, Andy likes to visit Karen Wall if he can. "Hall Farm to me always feels like being at home," he says. "Karen has been extraordinarily kind to me and my family." After a pause, he adds, "The other day, I was looking out from the churchyard at Countisbury, and it struck me that most of the land I could see had belonged to my family at one time. It's something we'll never get back, and that made me pretty emotional."

Recently, though, he has bought a small piece of land near his house in Crediton – the first land he has ever owned. "Some people don't grasp the pull a piece of land can have on you. It's so special, and it's also special what you do with it. There's a primeval desire to produce food from land, I think. Food should definitely be regarded as a public good."

Above: Andy with Robin May at Blackmoor Gate Market.

Above: Andy chatting to Robert Crocombe at Cutcombe Market.

Left: Andy in Cutcombe Market Cafe.

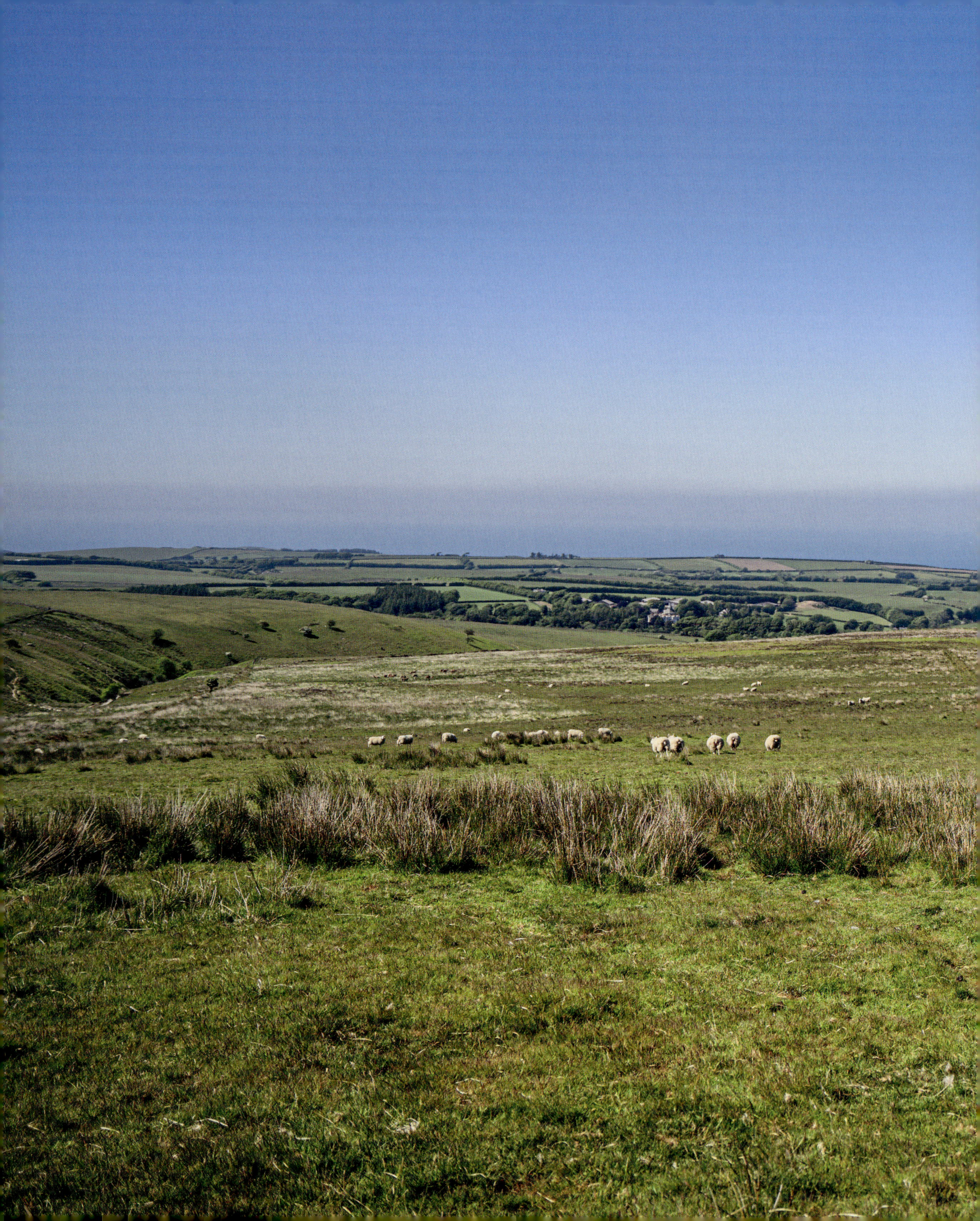

Woolhanger from Butter Hill.

The Woolhanger Farming Partnership
Parracombe

In 1972, Graham and Beverley Mellstrom bought Woolhanger Farmhouse and 600 acres. Since then, much more land has been acquired along with Woolhanger Manor and Music Room, which have been renovated.

The Estate is now owned by Beverley Mellstrom in partnership with daughter Juliet Slot and son Steve Mellstrom. Juliet is the managing partner.

The land farmed amounts to around 10,000 acres, including shared common grazing, a farm business tenancy and almost 1,000 acres grazed under licence on The Chains.

Livestock numbers have been reduced since the farm became organic about 15 years ago. There are now approximately 1,500 Cheviot ewes that lamb outside in three batches from March to May, 400 cows that calve outside from mid-April until the end of May, 85 pedigree Galloway cows that calve outside in the autumn, 550 young cattle and 17 bulls (Galloway, Charolais, Angus, Limousin, British Blue and Whitebred Shorthorn). There are also six pedigree Exmoor ponies.

"We've got a really good community here," Juliet says. "We've found the farm works best when we employ local people. Faye, our secretary, was only twenty when we took her on, but she's very bright and learned quickly. The other day we had a Rural Payments Agency inspection, and she did brilliantly."

Faye Reed grew up on Hunniwins Farm, near North Molton. "This is such a nice place to be. My dog Bella is my office buddy, and I can ride before I go to work," she says. "I was amazed by the scale of the place at first. There was a huge amount to learn, but the Exmoor Hill Farming Network's Bookkeeping and Mastering Medicines courses were a great help."

Technology assists Faye to keep detailed records. The Estate uses Shearwell Data's equipment as well as its 'FarmWorks' software, and the staff send Faye information by mobile phone as they record it – much more efficient than retrieving soggy scraps of paper from coat pockets at the end of the day!

A new management tool, virtual fencing, has been funded through the Exmoor Mires Partnership. The system allows invisible virtual fences to be drawn using a mobile phone. The app then communicates with GPS-enabled electronic collars worn by all the adult cattle within that enclosed space so they can be controlled, moved and monitored. The boundary can be moved easily, and cattle can be excluded from dangerous areas or places that need

Below: Cows and their newborn calves.

protection. Also, the movements of the cattle can be monitored.

"It's a useful piece of kit, but so expensive that we probably wouldn't have invested in it without a grant," says the farm manager, Sam Smyth.

Sam was brought up on his grandparents' dairy farm near Parracombe. "I've always loved farming, but I felt a bit trapped before," he says. "The job of herdsman at Woolhanger came at just the right time, and the manager's position followed much sooner than expected, when I was twenty-eight."

Five years later, he has already made changes. "There were about three hundred-and-fifty Charolais cattle here, which were a nightmare to calve outside. There's a huge risk of disease calving that many in a shed, so we've been replacing them with Angus and Bluegrey (Galloway cross Whitebred Shorthorn) cows, which are much hardier and cope well with calving outside," he explains. "And the sheep used to be Scotchies and Scotch Halfbreds, but we've bought good-quality Cheviots instead."

They have to jump through a lot of hoops to be organic, but they've got the space and staff to make it work. "We lay up about a thousand acres for hay and silage every year. Even so, I always worry about not having enough forage," Sam says. "TB is the massive thing that rules the business, though. Testing takes six days twice a year, and that doesn't include the time spent getting all the cattle back. We're in and out of TB restrictions the whole time. It's the most stressful thing about managing Woolhanger, but otherwise the job is very rewarding."

Sam's father Gary has joined the team as the Estate's tractor driver. "Working for my son is a natural progression – it seems to work alright," he says, smiling. "There's less stress and more free time than when we were on the family farm. I used to be stuck in the yard dairying, but now I get a different view every day and see lots of wildlife. This job has made me fall in love with Exmoor."

From top to bottom of page:

Sam and Faye studying a map of the Estate in the office.

Gary giving a field a spring makeover by aerating the soil and picking up stones.

Martin checking the cattle.

Tony rounding up some sheep.

Above: Cows and calves coming home to the farmyard.

Martin Petherick agrees. "I've seen more of Exmoor since I've been working here as herdsman than ever before," he admits. He grew up on his family's farm near Parracombe, was in the same year as Sam at school and was best man at Sam and Leanne's wedding. "It's lovely working here and, having been self-employed before, it's good to have a stable wage," he says.

Tony Leworthy is 54. He has spent a lot of his working life at Woolhanger, and has seen tremendous changes. "Back in the days of production subsidies, we grazed a thousand Welsh Mountain sheep all year round on Lynton Common alone and we didn't get many lambs from them," he says. "Going for quality rather than quantity and farming organically has helped the stock and the land. I've especially enjoyed the part I've played in improving the sheep. I love shepherding – there's nothing better than working with good sheep and good sheepdogs."

"The staff are key to making this work as an estate," Juliet concludes. "I'm not based here, but I know I can trust Sam and everyone else to do a great job. We're always looking at how we can improve things; my top priority is reducing our energy costs and environmental impact."

Later in the year, about 1,500 acres near Challacombe were transferred to Beverley Mellstrom's daughter Lindsey Roberts, who moved back from New Zealand with her husband Mark. Tony is now working for them at Challacombe.

Farm Plastic Recycling Scheme

Farmers are becoming increasingly aware of the environmental impact of their businesses. While the use of plastic has revolutionised every aspect of life, including farming, disposing of plastic waste can be a huge problem.

The Farming and Wildlife Advisory Group SouthWest (FWAG SW) started its farm Plastic Recycling Scheme in 2004, and since then the annual collection date has become an important fixture in the diaries of many farmers. There are two collection points on Exmoor: Cutcombe Market and Blackmoor Gate Market.

Below: Unloading plastic from a farm trailer at Blackmoor Gate. *(CE)*

Above: Sorting the different types of plastic. *(CE)*

The aim of the scheme is to recycle as much plastic as possible to prevent it from piling up on farms, going to landfill or being burnt. The scheme is simple and straightforward, with no membership fees so that farmers only pay for what they bring.

The types of plastic collected include silage wrap and clamp sheets, fertiliser bags and dumpy bags, feed bags, string, netting, chemical containers and mineral buckets, and tree guards.

The plastic is taken away, washed and melted down into pellets. The pellets are then recast into an array of agricultural products, such as animal housing, step-through stiles, cow tracks and eco ground grids, sheep pens and fence posts.

In May 2023, a total of 54 farmers went to Cutcombe Market with 33.6 tonnes of plastic for recycling, and in addition 34 farmers took just over 20 tonnes to Blackmoor Gate.

Lauren Clarke, Senior Farm Environment Adviser at FWAG SW, organises the scheme. "Collections take place at this time of year so farmers can have a spring clean after their livestock are turned out," she says. "There's always a buzz on farm plastic days, with farmers arriving a bit like buses – either none or five at time!"

Below: Baling silage wrap and other plastic waste on-site so it can be transported from Blackmoor Gate for recycling. *(CE)*

David Brewer and Kate Hughes
Wood Advent Farm, Roadwater

David Brewer and his wife Kate Hughes live at Wood Advent Farm with their two children. Kate's book *Going Zero: One Family's Journey to Zero Waste and A Greener Lifestyle* describes how and why they have worked towards having a zero-waste household, so it was a natural extension to try to farm without using plastic when they took over the business from David's parents, John and Diana, two years ago.

"We'd like to farm without plastics or fossil fuels, but it's nigh on impossible at the moment," David comments. "When I ask suppliers about their use of plastic, I get the impression they think I'm just being awkward. Other farmers need to start asking questions about this so it's taken seriously. We'd pay double for plant-based silage wrap, for instance."

They are still trying their best, with no fertiliser bags, no plastic buckets and using a weather-weave mat that lasts for about five years for outside storage. They can't avoid diesel, but David says he thinks twice before getting into a tractor.

Their grassland became organic in the 1990s and the arable land followed suit two years ago.

"We have four key criteria: soil health, the production of nutritious toxin-free food, biodiversity recovery and climate change mitigation," Kate explains. "In essence, it's a traditional rural lifestyle."

Below: David and Kate planting trees with their children. *(Oxygen Conservation)*

Above: Aerial view of Wood Advent Farm. *(Oxygen Conservation)*

They are aiming for a balance between food production and wildlife throughout the farm. Around 150 acres of agroforestry and 150 acres of woodland are being created, and they are also reinstating an old orchard and creating a food forest. A herd of pedigree Devons produce both breeding stock and meat, and additional store cattle are bought in for finishing if there's a surplus of grass.

On the better ground, carefully chosen arable crops such as naked oats and YQ (Yield/Quality) wheat are grown. Naked oats thresh free from the husk and make an excellent break crop, while YQ wheat has hundreds of lines crossed from 20 reliable wheats to increase resilience and genetic diversity.

"Coming into farming from a career in journalism and broadcasting has made it easier to have a different approach, and we're lucky that John and Diana are incredibly supportive about what we're trying to achieve," Kate says.

"We're constantly asking questions and reassessing what we're doing," David adds. "As well as knowing what we want to do, we've got to be sure why we're doing it. I find the present farm support schemes quite restricting – wildlife and farming are driven by the weather, not dates. We try to choose schemes to complement what we want to do rather than letting agreements shape our farm management. The regenerative farming movement is growing tremendously; hopefully it will be recognised and rewarded in future."

Kate is now working on a book that explores the principles behind what they're doing at Wood Advent Farm, which will incorporate issues such as food security and what we can do to take control of our own futures. She has found that the changes they've made so far have made their lives simpler, more fulfilling and less expensive.

Who wouldn't want that?

June

This was the UK's warmest June on record, and May's dry spell persisted until it eventually broke down with thunderstorms and heavy rain towards the end of the month.

Dry weather is ideal for sheep shearing, and June is when most sheep are shorn on Exmoor.

Oliver Hill and Kate South are two people who combine shearing in the summer with other farm work during the rest of the year. Oliver also farms with his brother Adam, and they run a contracting business together. Another enterprising person who has built up his own contracting business is Charlie Rook from Porlock.

As the Manager of British Wool's South Molton Grading Depot explains, several factors have contributed to a decline in the value of wool. Nowadays, it generally costs more for a sheep to be shorn than the fleece is worth. This is one of the reasons why some farmers, like Dave Knight, are breeding wool-shedding sheep.

June is also the month when many red deer have their calves. Julian and April Westcott farm red deer, so this is an important time of the year for them.

Above: Adam (left) and Oliver (right) with their David Brown tractors.

Adam and Oliver Hill
Well Farm Bungalow, Timberscombe

Adam and Oliver Hill have lived at Well Farm Bungalow, near Timberscombe, since the age of two, when their father and mother, Wilfred and Philippa Hill, moved from Smallacombe Farm near West Anstey.

The boys loved farming from an early age – a passion fuelled by a neighbouring farmer. "I don't know whether we should thank him or curse him, because we became farming mad and used to 'help' him as much as possible," Adam says.

At home, their Britains farm toys were treasured possessions, bought with any savings they had. "We haven't changed much," Oliver jokes. "Our toys have just got bigger, that's all."

Their father died at the age of 82, when the twins were nearly 14 and their younger sister was 10. It was a huge shock for all the family.

Soon afterwards, the boys began their farming careers with the purchase of six ewes. "The next year we had twelve sheep, and we've gone on from there," says Oliver.

The brothers owe a huge amount to local Young Farmers' Clubs: first Cutcombe and then Kingsbrompton, based in Brompton Regis. They made some lifelong friends and learned many skills, including public speaking, stock judging, hedgelaying and shearing. Through Young Farmers, they both went on a British Wool Board shearing course hosted by Shearwell at Cutcombe.

"That's where Oliver got the shearing bug," Adam comments. "I can shear a sheep if I need to, but I wouldn't want to do hundreds like my brother."

At West Somerset College in Minehead, they both chose Agriculture as an A level subject, and they went to different dairy farms for work experience. Adam was offered a job three days a week after he left college, but Oliver had done enough to know he didn't want to be a dairy farmer. He worked for six months on a mixed beef and sheep farm in Exebridge before doing some pre-lambing docking on various farms, followed by his usual lambing work near Wheddon Cross. After that, he did wool tying for another shearing contractor and ended up

Above: Oliver shearing an Exmoor Horn ewe at Yarner.

shearing a few sheep, too. "Thirteen years ago, I sheared my first hundred in a day, and got hooked," he says. "I sheared my first two hundred at Great Nurcott Farm, and to celebrate the occasion they played AC/DC's *Highway to Hell*. I remember it whenever I hear that song."

At the same time, the brothers were farming their own sheep and doing some work with an old David Brown tractor they bought in 2011 from Charlie Rook. They entered the 2012 Exmoor Society Pinnacle Award with their idea for a small agricultural contracting business, called A&O Hill Agricultural Services, and won it. "We bought a mower and a baler, as well as a few other bits of machinery, which put us forward quite a few years," says Adam.

The following year, the twins helped out at Lyncombe Farm, near Brompton Regis, because their friend Jack's father was ill. Adam has carried on working on the farm one day a week, and he's been working part time there ever since.

Above: Oliver rowing up hay, ready for baling.

Above: Adam baling hay. The Hills' machinery is ideal for small fields with tight corners.

In 2019, Adam gave up his dairying job and saw an advertisement, through the EHFN, for help wanted on a traditional beef and sheep farm near Exford. Now he works part time on that farm and at Brompton Regis – as well as working with Oliver, looking after their own sheep, doing contract work and fixing the odd bit of machinery. "I often get home after five and start work all over again," he says.

Oliver has been contract shearing for 10 years now, and a few years ago he started Hillbilly Haircuts. "It's my little concern, but a few people help me out," he says. "We must be the most relaxed shearing gang on Exmoor – we do a good job and enjoy what we do." This year, Jack Bishop, Dan Matravers and Callum Ash have been shearing with him, and Nathan Brooks is also helping.

"I do like sheep, I don't know why," Oliver says. "I really enjoy shearing the Exmoor Horns at Yarner. John Richards has got the best large flock of them I've ever seen. He's a good stockman – very particular. Some shearers don't like Exmoor Horns, but I don't know what all the fuss is about; you've just got to hold them differently."

He also enjoys going to Wydon Farm to shear any wool-shedding sheep that haven't shed all their wool. "They're good fun to shear; it just melts off."

Below: Draying straw over Landacre Bridge. *(Adam Hill)*

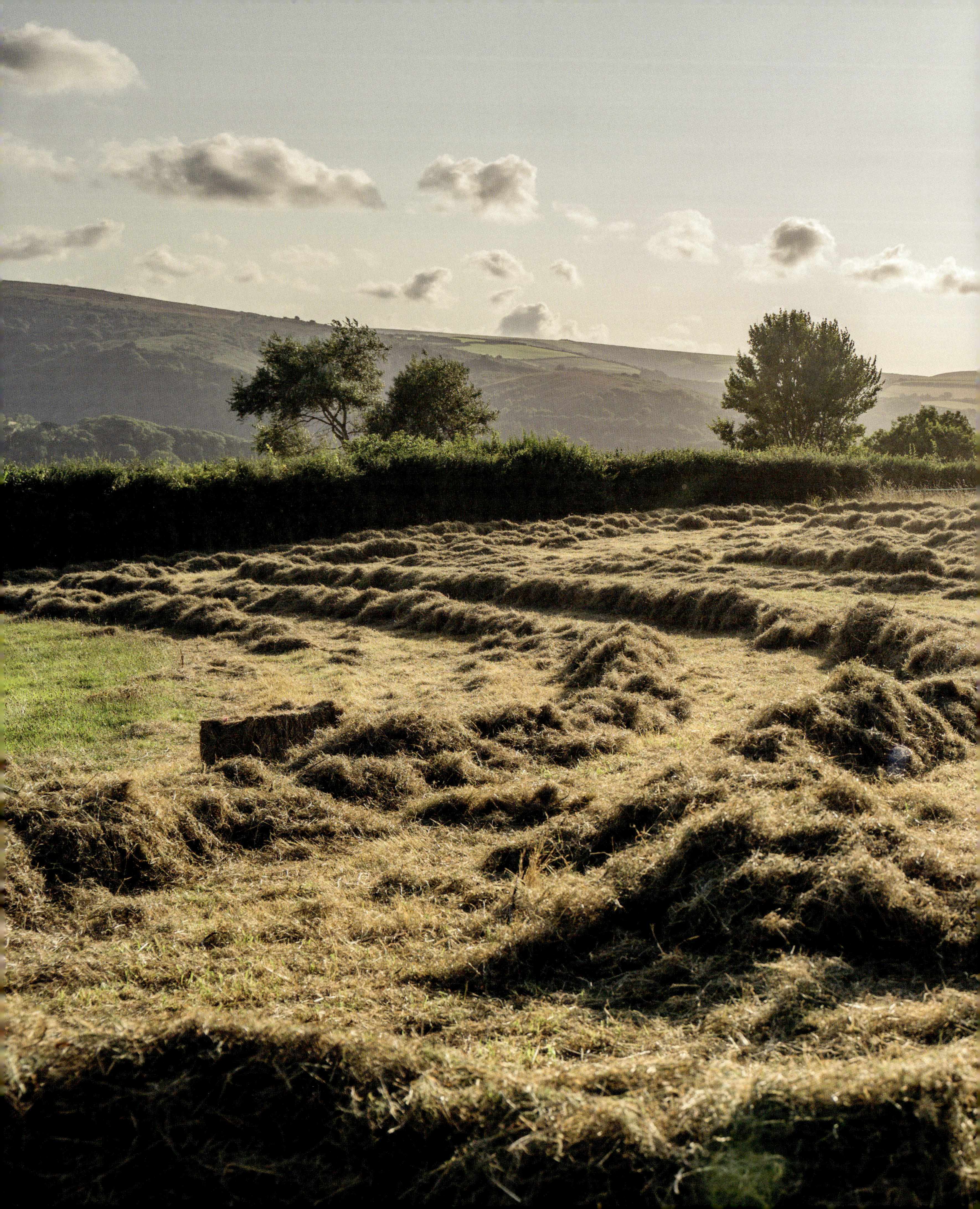

Haymaking at Selworthy

Above: Charlie with David Greenwood at Cloutsham, doing groundworks for a new cattle handling facility. (EHFN)

Greenwood at Cloutsham and Richard Richards from Silcombe. Soon he had a mower as well, and he bought a square baler and bale wrapper from Geoff Tucker, which set him up for grass harvesting work in the summer.

"The trouble with hay and silage is they're so weather dependent. Most people around here do a single cut, so if it's a catchy summer about three-quarters of the workload comes in two or three weeks," Charlie says. "My main focus is on digger work and groundworks. I was really pleased when Peter Huntley asked me to do the groundworks for the new development at Blackmoor Gate – it's nice to still have that connection with the Market."

Charlie also does work for the NT, ENPA and various private landowners, including about a month a year at The Lillycombe Estate, near Porlock. He has tree shears and does hedging in the winter for people, too.

Ollie Curtis works for him part time. Although Ollie's self-employed, Charlie still feels a responsibility to bring in enough work. "It seems to be all or nothing with farming, but I'm lucky with the people I work for. I haven't had to advertise yet."

His younger brother George sometimes works with him, and his dad helps with the mechanical side of things. They don't have sheep on their fields anymore, but lay them up early and crop them for hay, which they sell. In addition, they buy standing grass from people and make hay from that, which they store in their shed.

"I'm happy with what I'm doing and want to stick with what I know. I don't really want the business to get any bigger, otherwise the hassle could outweigh the enjoyment," Charlie says. "It's satisfying doing a good job, and nice when you can make somebody's life a bit easier."

Kate South
Farley Water Farm, Brendon

"I've always loved farming," Kate says. "I did try to be a townie for a while when I was a teenager, to fit in with my friends from school. I even caught the bus into Barnstaple on Saturday mornings to wander around the shops with my friends, carrying my little handbag, but it didn't work very well!"

Kate's home is Farley Water Farm, near Brendon, and most of her school holidays were spent farming in one way or another. Her dad Ian, who was one of the first men on Exmoor to shear 400 sheep in a day, taught her how to shear when she was a teenager, and she enjoyed it so much that she wanted to do it for a living. However, other people told her it wasn't a woman's job, so she abandoned the idea, left school after doing her A levels and, when she was 20, started working for Torch Farm Vets, where she was a veterinary technician for the South West Sheep Breeding Team.

Six years ago, Kate met Fred Jones. At the time, he was working for John Tucker at Stetfold Rocks Farm, near Exford, but now he's a sheep shearer. 'If you want to shear, then shear,' he told her, so she started shearing part time during the season for a few years. Then, in 2022, she decided to give up her veterinary work and travel to New Zealand for the sheep shearing season there.

"New Zealand is beautiful – there's no other way to describe it – and the people are lovely, too," says Kate. "The towns are more like villages, and everyone seems to be involved with farming in one way or another. The stations are huge, with lots of workers – a real community thing."

Kate was with a gang in the Hunterville area of North Island. She often had to leave at five in the morning to get to work in good time because shearing always started on the dot of seven. "It was quite relentless: eight shearers

Below: Kate with her dad Ian, shearing at Cheriton. *(Chris Allen)*

and ten thousand plus sheep. We did two-hour stints four times a day, and we all changed our cutters every quarter of an hour."

She had a wonderful time in New Zealand, and learned a lot, but by the last week or so she was ready to come home. "I'm still an Exmoor girl through and through."

Once home, she was straight into a particularly difficult lambing at Farley, and then, on 10th May, she started shearing full time in the Exmoor area with Adrian Kingdon's gang.

"Adrian's a very good boss; nothing fazes him," she comments. "I'm with Will Edwards and Chris Ley, so we're the next generation. Before that it was Michael Kingdon, Geoffrey Ley and Dad."

Kate says that although it's hard not to be competitive sometimes, she's not one for racing too much. "The main thing is to do a good job and keep the sheep calm. It's a self-preservation thing as well, because my body's got to last the whole season. Swimming in the sea after work does me the world of good. We're so lucky to live near the coast."

There's only been one day in the past month when Kate hasn't held a handpiece, but she's really enjoying her first shearing season on Exmoor. "It's quite daunting being self-employed. Apart from helping Dad at home, I'm not sure what I'll be doing when the season ends," she admits. "Work's always available if you want to do it, though."

Above: Kate shearing a Scotch Blackface on the Simonsbath Estate.
Below: Shearing at Simonsbath on a specially designed trailer. L-R: Kevin Harris, Kate South, Adrian Kingdon.

Exmoor Horn Wool

Several years ago, Lindy Head had an Australian bed and breakfast guest who was on a mission to collect knitting wool from every British native breed of sheep. Despite being an Exmoor Horn breeder, Lindy had to admit she didn't know of anyone producing Exmoor Horn knitting wool, but it got her thinking.

She talked to John Arbon, who had a spinning mill in South Molton producing both knitting wool and socks and, encouraged by his enthusiasm, a meeting of the Exmoor Horn Sheep Breeders' Society (EHSBS) was arranged to gauge how much interest there would be in producing knitting wool and ready-made items like socks using fleeces from Exmoor Horns.

"John Arbon provided the impetus that got us going," Lindy remembers. "After that it was a steep learning curve. Mastering the technicalities of what happens to wool was really hard, but not nearly as hard as selling, which involved a whole new skill set."

Thanks to the sheer hard work of Lindy and other dedicated committee members, the official launch of Exmoor Horn Wool Ltd took place in October 2015.

By that time, both knitting wool and knee-length socks were available to sell, and the company was promoting Exmoor Horns and their wool at events like 'Woolfest' in Cumbria and 'Yarndale' in Yorkshire.

"We chose to produce knee-length socks to begin with because there was a huge local market for shooting socks and the wool from Exmoor Horns is particularly hardwearing," Lindy says. "After that we developed calf-length walking socks, posh socks, pullovers, Fair Isle knitting patterns and kits, cushions and a throw. Our walking socks have been the greatest success – they're the most popular things at Christmas, along with our hat kits."

Because all the energy-intensive processes involved in the production wool, like scouring and spinning, have become much more expensive recently, the manufacture of more stock has temporarily been put on hold. However, Exmoor Horn Wool continues to promote the breed and its wool, as well as exploring new avenues.

For instance, in 2021 Vispring created a luxury mattress using Exmoor Horn fleeces, giving a generous donation, equivalent to 50p for every kilogram it used, to EHSBS members in proportion to their contribution towards the total clip delivered to the South Molton Depot. Exmoor Horn Wool facilitated this. Dick Tucker was chairman of the EHSBS at the time, and Vispring made a film of his farm with a commentary by Kate Humble, which has raised the profile of Exmoor Horns.

Above: Lindy with some of her Exmoor Horn sheep. *(Emily Whitfield-Wicks/NFU)*

More recently, ten million pounds have been made available nationwide through the Coronation Fund to support native breeds and, along with gene banking, some of that may go towards promoting wool from native breeds of sheep now that the Wool Board can track each clip.

"You never know what's around the corner," Lindy says. "We need to make the most of every opportunity that comes along."

Right: Julian Branfield promoting Exmoor Horn Wool at Dunster Yarn Market by giving a demonstration of shearing using hand shears. *(Lindy Head)*

Below: Hector the ram manning (or ramming) the Exmoor Horn Wool stand at a wool festival in Cumbria. *(Lindy Head)*

Royal Cornwall Show

Clockwise from top left: Richard Clark with Closewool sheep belonging to his wife's family; Jim Pile, from Kentisbury, exhibiting his Border Leicester sheep; feeding time for two prize-winning Exmoor Horn rams; Reserve Male Champion, Ralphy, bred and shown by Dick Tarr of Churchtown Farm, West Anstey.

The South Molton Grading Depot

British Wool is the only organisation in the world that collects, grades, sells and promotes fleece wool. Established in 1950, it is the last agricultural commodity board in the UK and is a farmer-led, non-profit organisation.

It has three grading depots in England: Ashford in Kent, Bradford in West Yorkshire and South Molton on the southern edge of Exmoor.

In pre-industrial times, wool was the most valuable commodity produced on Exmoor, and most of the market towns in the area were founded on the wealth it generated. Sadly, many factors have led to a decline in the value of wool over the past hundred years or so, including the development of synthetic fibres, shifting global markets and, in 1992, the end of the government price guarantee.

Wool now makes up just one per cent of the global fibre market, and British wool is only two per cent of that.

In the UK, it usually costs more to get a sheep shorn than the fleece is worth.

"Our mission is to maximise the wool value for our members, but it's sometimes a tough job persuading farmers we're on the same team," says Adrian Prisk, Manager of the South Molton Grading Depot. "Wool is an incredibly bulky commodity, and it is a surprisingly long, expensive process to get to the end product. Scouring costs alone have gone up thirty per cent since the recent energy crisis."

Nationwide, about 23,500,000kg of wool were collected last year – all graded by hand.

Below: Inside the depot, a wool sack containing 20 or more fleeces is about to be emptied.

Above: Grading the fleeces.

"You need touch and feel for grading," Adrian explains. "This is the only country in the world where grading is so specific, with over a hundred different types. It's a high-pressure job requiring both skill and speed, which come with experience. The target weight of wool to grade in a day is five tonnes per grader. That's okay if the wool is, say, all Exmoor or Romney, but most farmers tend to have several types of sheep. A lot of different breeds in the same bag is a nightmare, and crossbreds are often especially difficult to sort."

Typical auction lots are just over eight tonnes of a single type of wool, so one lot often contains wool from several producers.

At the moment, there are four main graders at South Molton: Alan, Charlie, Dave and Sam. Adrian also grades when needed, and Richard is a trainee in the first year of his apprenticeship. They work every weekday from the beginning of May until the end of March. Holidays are taken in April, and maintenance jobs also take place then.

"Andrew Hogley, our CEO, is doing great things like promoting a new e-commerce website and traceability so that niche markets can be found," Adrian says. "A few years ago there was hardly any premium for organic wool, for instance, but we were able to pay a pound per kilo extra last year. And there's increasing demand for wool from certain breeds, like Exmoor Horn and Zwartbles. We're also working hard to support new products, like biodegradable tree guards. As the long-term environmental problems associated with plastic and synthetic fibres are realised, the future should be much brighter for wool."

Dave Knight
Wydon Farm, Minehead

Dave Knight was Chairman of the Exmoor Hill Farming Network from 2014 until 2022, when he stepped down due to increasing family commitments.

He was born and raised at Wydon Farm, near Minehead, which he farms with his father, his wife Jenny and brother Sam. They rent about 700 acres from the NT and a private landlord, own 200 acres and also graze various areas for the NT when required. The land ranges from fields on the outskirts of Minehead, at about two hundred feet, to moorland above the sea cliffs at nearly a thousand feet.

For the past fifteen years, Dave has been building up a flock of sheep that don't need shearing based on the Exlana.

"I remember reading an article in *Farmers Weekly* about wool-shedding sheep nearly twenty years ago," Dave says. "Dad and I both thought it was ridiculously funny – fancy breeding sheep that shed their wool all over the place!"

Above: Dave on his quad bike, which is equipped with electric fencing materials for rotational grazing. *(EHFN)*

At the time, the majority of their sheep were Exmoor Mules, and every year they lost several that had become stuck on their backs or entangled in brambles.

Below: Cattle at Wydon Farm. *(EHFN)*

Above: Dave's wool-shedding sheep are particularly good at grazing the unimproved pastures along the coast. *(Shaun Davey)*

"One day, after finding yet another ewe cast on her back, I remembered that article and had a lightbulb moment. Hang on, I thought, I bet wool-shedding sheep wouldn't get stuck on their backs or caught in brambles."

As luck would have it, 20 Wiltshire Horn ewes were advertised for sale near Braunton, so Dave bought them, just as an experiment, to put with a Wiltshire Horn ram.

"I liked the wool-shedding but didn't like their temperaments; everything they did was awkward," he recalls. "Then one day I was at Peter Baber's farm near Exeter. He was one of the farmers who developed the Exlana breed, and I ended up buying an Exlana ram from him."

Dave put the ram with some of his Exmoor Mule ewes, and he freely admits he wondered what on earth he'd done because most of the first cross ewes didn't shed properly and looked so untidy. However, the second crosses shed much better and almost all of the third crosses shed completely.

"About seventy to eighty per cent of the flock are shedders now. It's been a long process, but I'm nearly there," Dave says. "I'd like to have been able to speed everything up by buying in pure-bred Exlanas, but the price has rocketed."

Wool-shedding sheep have a tight coat that dries out quickly, and they moult gradually when they're ready in the spring or early summer. "It's a misconception they can't handle bad weather," Dave remarks. "I lamb my ewes outside, and they do really well. Also, wool that's shed on the fields feeds the soil. Birds love it for their nests, too."

All things considered, Dave says he's glad he stuck with his plan to create a wool-shedding flock. Soon none of his ewes will need shearing or docking, which will save both time and money.

"The shedders are exceptionally good at grazing the rough coastal ground. They eat whatever they can find, and they range all over the place," Dave says. "In fact, the only downside I can think of is that the dogs find gathering much more difficult now!"

Julian and April Westcott
Mill Reef Farm, Roadwater

When Julian Westcott was 21, he visited New Zealand and was inspired to try deer farming on his return to Exmoor. His father let him have one-and-a-half acres, and he bought seven fallow deer from Powderham Castle.

Julian soon switched to red deer, because they're larger and have better conformation, and within three years he was farming 60 acres of land at home, with six-feet-high fencing and a herd of English Park red deer with Furzeland and Warnham bloodlines. Park deer are different from wild red deer due to hundreds of years of selection for desirable traits.

Before long, Julian was joined by his wife April, who has worked alongside him on the farm ever since. "It was early days for the deer farming industry here," he says. "April and I teamed up with a local abattoir and a national supermarket for a trial run. Other deer farmers joined in, and soon there were thirty-five of us supplying outlets, from Cornwall to South Wales and Oxford."

Their two daughters, Abbie and Robyn, are also involved in the farm. They now have a closed herd of 80 hinds and followers plus eight stags.

"We select for temperament first and foremost, followed by conformation, health and also antlers," Julian says. "Some of our deer are twentieth generation now. It's important that the deer are easy to handle to minimise stress."

Above: This stag is 'in velvet'. Farmed deer often have their antlers removed for safety. In the wild, red deer on Exmoor typically cast their antlers from March to May, and grow a new set – covered in furry skin, or velvet – from May until August.

"The life expectancy of the hinds is around 14 years, and one of them lived to be 26, so we get to know each member of the herd well," April adds. "They're part of our family – part of who we are. The children have grown up around them. We all love the deer and deer farming."

The Westcotts are rightly proud of the animal welfare standards on the farm. To minimise stress, both prime deer and cull animals destined for meat are killed while they're still in the field.

Male calves not chosen for breeding are killed for meat from around 10 months old to produce a year-round supply. The farm is a licensed farmed game-handling facility, which means it isn't governed by the game season (unlike Park deer and wild deer). The animals are inspected by a vet and the carcasses are refrigerated before being delivered to local processors who have ordered venison.

Left: Julian feeding some red deer hinds. They moult during May and June.

Above: A hind giving birth.

"Farmed venison is different from wild venison," April says. "It's not gamey and doesn't need hanging." She is training as a butcher and works for a local butcher, abattoir and retail outlet.

The Westcotts' prizewinning herd has a very good reputation within the industry, so finding good breeding homes for surplus hinds isn't a problem. They have even sent live deer, embryos and semen to New Zealand.

By the beginning of September, the stags are out of velvet and their antlers are cut off to prevent injuries. This doesn't hurt them; it's done every year because stags naturally shed their antlers in the spring and grow a new set.

The hinds are about 15 months old when they are mated. Two stags are put in with each group of hinds early so they can all get to know each other.

"It's important for young stags to have a few seasons of being an underdog as it teaches them manners and prevents them from becoming hard to handle," Julian explains. "Having two stags insures against fertility problems, too."

The stags and hinds are separated at the beginning of November, and the deer are housed in the first week of December. This is when weaning takes place as well.

After a seven-and-a-half month gestation period, the hinds will drop their calves in May or June, and so another production year begins.

Above: A newborn red deer calf hiding in the long grass.

July

The focus of farmers at this time of year is usually on conserving enough forage for the winter, but July was unusually cool and wet, which delayed a lot of hay and silage making.

Seeds from wild flowers are also harvested in July, notably through the ENPA's Sowing the Seeds project. This year, the ES organised a Meadows Day in collaboration with the ENPA, FWAG SW and the Exmoor Forest Estate, where guests saw some beautiful wild-flower meadows and learned about their management.

The Meadows Day was just one of many collaborative events involving farmers and conservationists that take place throughout the year on Exmoor.

A few miles away, Serena and Mike Colwill welcome visitors for educational visits to Coombe Farm, from primary school children to fellow farmers.

Jack Buckingham is another innovative farmer who is putting new ideas into practice on his family's farm. He is also a farm business consultant.

There are always challenges to overcome and new things to learn when you are farming.

The Exmoor Forest Estate
Simonsbath

The Exmoor Forest is the historical and geographical heart of Exmoor. The area became a Royal Forest in Saxon times – a designation formalised with Forest Law by the Normans. The word 'forest' in this case meant land reserved for hunting.

The ancient Royal Forest of Exmoor was enclosed by Act of Parliament in 1815, and by 1820 most of it (around 20,000 acres) had been purchased by a wealthy industrialist called John Knight. John and his son Frederic reclaimed a huge amount of moorland and built several farms and roads.

The 1815 Act had set aside 12 acres in Simonsbath for a church, vicarage and homestead, and 41 years later William Thornton became the first Curate of the Parish of Exmoor – the largest parish in Somerset.

With no direct heir, Frederic sold the Forest Estate to the Fortescues from Castle Hill, Filleigh, and it remained intact until 1959, when Lady Margaret Fortescue sold over 9,000 acres to pay death duties. Some of that land is now owned by the ENPA.

In 1995, after nearly 100 years of ownership by the Fortescue family, the Countess of Arran (Lady Margaret's daughter) sold the remaining part of the Estate to John Ewart – except Balewater, home to the Exmoor Foxhounds, which is still owned by the Fortescue family.

Duredon Farm, next to Balewater, was also excluded from the sale because it had been sold previously.

Mark de Wynter-Smith came to Simonsbath in 1997, to work as farm manager for John Ewart, and he has remained the Estate's farm manager since then.

John Ewart decided to sell in 2006, and Dr the Hon Gilbert Greenall bought the Estate.

In addition, Duredon was bought in 2017, and a new house – used for holiday guests – was built on the footprint of the original Knight farmstead.

In 2021, Ed Greenall (Gilbert Greenall's eldest son) bought the Exmoor Forest Inn, Simonsbath, with his brothers Freddie and Alexander. Including apprentices, 20 people are employed there, and beef, lamb, mutton and venison from the Estate can be found on the menu.

Above: Farm Manager Mark de Wynter-Smith with Ed Greenall.

The Exmoor Forest Estate is now about a third of the size it was when John Knight owned it, with just two farms based at Cornham and Simonsbath Barton. In total, it amounts to 5,866 acres, which includes 3,798 acres rented from the ENPA. Much of the land lies at over 1,000 feet, with the highest ground up to 1,500 feet.

Mark has seen huge changes since the mid-1990s, driven by government policies. "Headage payments caused a lot of damage, and their removal was a big positive," he says.

"There were five-and-a-half thousand ewes on the farm when I first came here, and we had to use three hundred acres for winter feeding. Those areas were black with mud – it was dreadful. And we went to over six hundred suckler cows."

Going organic in 2010 was also a big positive, not least because it gave them an extra incentive to reduce their livestock numbers.

The Estate now carries a herd of around 100 pedigree Galloway suckler cows and about 170 Angus and Beef Shorthorn cross cows.

"We are continually improving the breeding of our sheep as well," Mark says. "The main flock is of around thirteen hundred Lanark-type Scotch Blackface ewes. They seem to thrive here, and cope with the high rainfall very well. We also use a crossing type of Bluefaced Leicester on the Scotch ewes to get a good, dark Scotch Mule to sell."

Mark is glad he has tried to do what he has considered to be right, with or without subsidies. For instance, when the Estate was in an Environmentally Sensitive Area (ESA) scheme, he was told he should cut down the amount of low-input grassland on the farm, but he felt it would be a pity to do that because the biodiversity in those fields had improved. He continued to manage those fields to encourage a diversity of wild flowers, and they eventually became some of the foundation fields for the Sowing the Seeds project, which has proved to be a tremendous success. "A good relationship with bodies like the Exmoor National Park is dependent on a few key people, and both Heather [Harley] and Lucy [Cornwall] are really good."

Other initiatives haven't been quite so successful as far as he's concerned. "The Mires Restoration Project involved a lot of money and work, but I can't really see that the benefits have justified the majority of the work they did."

The Estate is now in a CS scheme, but the changeover wasn't straightforward – not least because the rules kept changing and some of the grazing prescriptions were far too restrictive.

"The proposed stocking calendar for the high ground between Prayway and Driver would have meant us reducing our stocking rates by ninety per cent, which wouldn't have done any good at all and wasn't viable as far as the farm was concerned, so we had to keep a lot of land out of an agreement and forego a large amount of money," Mark comments. "The people who make the rules don't listen to the people who know their land and livestock… I find it very insulting when environmentalists say Exmoor has suffered ecological decline in the past twenty years or so when I, together with many other farmers and land managers, have worked incredibly hard to improve biodiversity and make space for nature – and the results are plain to see. Sadly, all that sort of thing breaks down trust."

Mark rates TB testing as the most stressful aspect of his job. "We've had TB on and off for twenty-five years," he says. "The cattle have a whole herd test every six months. Winter testing is okay logistically because the cattle are all housed in sheds, but summer testing is a nightmare. It takes us ages to get them back from where they're grazing, and it's stressful for cattle and humans alike."

Another thing that has changed is the number of staff employed on the two farms. When Mark became farm manager, there were three shepherds, two stockmen and three tractor drivers. Now there's head stockman Paul Richards (aka Stompy), assistant stockman John Stapleford, shepherd Andy Pollard, his partner Janet Down (who helps with lambing and flock tasks throughout the year) and Jerry Coward, who does fencing and repairs.

Andy is from Porlock and Janet grew up in Simonsbath. They met hunting and have been together for 30 years. Andy has been shepherding on the Simonsbath Estate for 27 years.

"Janet's a great help, especially at lambing time," he says.

"We lamb from the middle of April, when the grass has begun to grow. You've got to farm with nature up here. Janet does until about seven at night and I finish around eight or so, and then there's a night lamber until we take over again at about six in the morning – that's when Janet gives me my orders for the day! Seriously though, you can't beat a good woman in the lambing shed."

He agrees with Mark that the sheep have improved dramatically since the Estate has been organic and there have been fewer animals to look after. "It's much more enjoyable producing well-cared-for, better-quality stock."

Most of Andy's spare time is spent hunting and point-to-pointing. He keeps a point-to-pointer and likes to support local race meetings. "The farming, hunting and racing community's all part and parcel of life on the moor, isn't it?" he says.

Stompy was also brought up in Porlock. His grandfather and uncle farmed Driver (near Simonsbath) and he spent a couple of years working there before working for other farmers. His first job on the Exmoor Forest Estate was lambing with Andy at Cornham in 1996. Various other jobs followed until he was employed full time as assistant

Below: Cattle in front of Cornham Farm.

Above: The Exmoor Forest Inn, Simonsbath.

herdsman in June 1999, working under Janet's father Gerald Down. A year later, Gerald retired and Stompy stepped up to his role.

"We used to calve over six hundred cows – indoors and outside – for twelve weeks, which was a massive task, and TB testing was a huge operation as well. It's much less stressful all round with fewer cattle nowadays, but TB testing is still the worst thing we all have to deal with," Stompy says. "Mark, Andy and I have been working together for twenty-five years, but it doesn't feel that long. We get on really well. Dr Greenall's a very nice guy – nice to work for – and Mark's a good manager. He trusts us to make our own decisions about the day-to-day things and involves us in big decisions, too. I've raised my family here – it's where I belong. I always wake up wanting to go to work, so I'm very lucky."

John grew up at Lucott Farm, Porlock, where his father was the stockman. At 20 years old, he went to work at Horsen Farm (near Simonsbath) for four years and then had jobs away from farming for a while as well as being a retained firefighter in Porlock Fire Service, where he became Crew Manager. His wife Donna grew up at Bromham Farm, Porlock.

"I heard the Exmoor Forest Estate may be looking for an assistant stockman, so I rang up Mark and got the job," John remembers. "I started in May 2018 and married Donna in August the same year. We're very happy here, with our two little girls Lily and Rosie. The job comes with a house, I've got a steady wage and I'm glad to be back farming again. I like Simonsbath, and I don't really mind the weather. Farming's the wrong job to be in if you worry about the weather."

Above: Meat from the Estate is served in the Exmoor Forest Inn.
Below: Steaks are especially popular.

Above: Head Stockman Paul Richards ('Stompy'). *(EHFN)*

Above: Shepherd Andy Pollard.

Below: Assistant Stockman John Stapleford.

65

Exmoor Sowing the Seeds Project

"It all started when some money became available for a conservation project through CareMoor," says Heather Harley, the ENPA's Conservation Officer. "Ali Hawkins and I both felt that Exmoor's grasslands were slipping through the net as far as funding was concerned, so we suggested the money should be put towards the restoration of wild-flower meadows."

They wanted to spend wisely, so they took advice from Simon Tomasso at the Devon Wildlife Trust. Thanks to him, they purchased a brush harvester, which allows seeds to be harvested without cutting the meadow so that hay or silage can still be taken from it later.

Heather and Ali also worked closely with Olly Edmonds from FWAG SW, together with South West Water and Natural England (NE).

In 2021, they harvested 60 kgs of species-rich seed. "We posted a tweet asking if landowners on Exmoor would like some seeds, and had an overwhelming response," Heather remembers. In 2022, they harvested 110kg seed, trained more people to use the harvester and engaged more land managers (24 farm businesses, 15 smallholdings and five community groups in all).

The Project had a tremendous boost when it received a Farming in Protected Landscapes (FiPL) grant that enabled a full-time project officer, Lucy Cornwall, to be employed for 14 months from February 2023. Her input has been so valuable that an application is being prepared to expand the Project in various ways, including research into the benefits of species-rich grasslands for livestock health and farm diversification. Heather and Lucy have been working with East Anstey School and Pinkery Outdoor Education Centre, and they would love to extend the educational aspects of the Project as well.

Above: Harvesting wild-flower seeds on the Exmoor Forest Estate. *(SE)*

They have been very busy this year, making new connections with landowners and land managers as well as organisations like Moor Meadows on Dartmoor, the Eden Project and, closer to home, The Exmoor Society. They're also hoping to inspire people of all ages with various meadows-related workshops: drawing, creative writing and even scything with a British scything champion.

There are now 58 landowners on the Sowing the Seeds database, and 160 kg of wild-flower seeds have been harvested in different places across Exmoor so far this year. Donor meadow owners get some money and, if they want them, seeds. Harvested seeds are spread on a tarpaulin in a dry shed for two or three weeks before being bagged into hessian sacks. "They need to be kept cool, dark and dry, and ideally should be sown by October the same year because some seeds, like yellow rattle, won't germinate if kept too long," Lucy explains.

Left: L-R: Lucy Cornwall and Heather Harley. *(SE)*

The seeds of specific plants are vacuum harvested for propagation in a wild-flower nursery where less vigorous species, like devil's-bit scabious, can be grown and plug plants can be cultivated to enrich existing meadows.

"We're learning all the time, and gathering good-quality data from the start is a vital part of that process," Lucy says. "People who want to create or enhance a wild-flower meadow have a site survey and are given a species list and soil report plus a bespoke management plan, as well as seeds to sow that are from a donor site with similar conditions. We also keep in touch afterwards to give practical support and monitor progress. More flowers mean more food, so we're seeing tremendous benefits for butterflies and biodiversity in general. Bird's-foot trefoil, for instance, can be a food source for around 130 different species of invertebrate."

"Yes, it's such a joy to visit meadows afterwards – very rewarding," Heather agrees. "It's a real privilege to visit farms and, through the Project, help people to enhance their meadows at any scale. Generations of livestock farmers have created Exmoor's wonderful species-rich places. Grazing livestock are absolutely essential, producing high-quality food and high-quality habitats. I'm not just saying this because it's my job; I'm truly passionate about how critically important the farming families of Exmoor are to the National Park."

Above left: Harvested material ready for processing. *(SE)*

Above right: Unloading the harvester. *(SE)*

Left: Tim Parish and Kate Lacey sieving wild-flower seeds. *(SE)*

Above: Mark de Wynter-Smith talking about the management of the Estate's wild-flower meadows. *(VE)*

Meadows Day at Simonsbath

The first ever Exmoor Meadows Day was held in Simonsbath at the beginning of July, organised by the ES in collaboration with ENPA and FWAG SW. The morning and afternoon sessions started with a talk by Olly Edmonds from FWAG in White Rock Cottage (owned by the Simonsbath and Exmoor Heritage Trust) followed by a walk into the wild-flower meadows of the Exmoor Forest Estate. This was led by Mark de Wynter-Smith, together with Heather Harley and Lucy Cornwall from the ENPA's Sowing the Seeds Project.

Mark talked about how the Estate's wild-flower meadows have developed since they have been managed on a low input organic system: grazing by sheep through the winter and then no grazing for about five months until a crop of hay or silage is taken in late summer, after which cattle graze for a few weeks before the sheep go in for the winter again. During the period when the plants are left to grow and flower, seed is harvested using the ENPA's brush harvester. Mark said it's important to take the cut herbage away so soil nutrients are kept in check and vigorous grasses and other plants like docks don't smother the slower-growing wild flowers.

Many people were amazed by what could be achieved through grazing management alone, without sowing any additional seeds, and the meadows were buzzing with life.

Serena and Mike Colwill
Coombe Farm, Exford

Farming, children and education have been strong themes running through Serena Colwill's life. She grew up at Coombe Farm, Exford. Her mother used to work at Red Deer Nursery, near Exford, and as a teenager Serena helped out there, too, gaining valuable experience for a NVQ in Childcare, which led to a job as a preschool leader at East Anstey. Serena's younger sister lives near Bristol and teaches Biology, so teaching is definitely in the genes.

Like many farmers, Serena and her husband Mike met through Young Farmers. When they were first married they rented a farm near Holsworthy, and they had triplets when Serena was just 23 years old. "We were young and energetic then!" Serena says.

Ten years later, their energy still seems to be boundless.

In 2020, Serena's parents, Michael and Denise Stanbury, moved to Wiveliscombe and passed Coombe Farm on to her. The Colwills wasted no time in setting up a campsite for 28 days during August to cater for increased demand due to the Covid epidemic, and they added cattle to what had previously been a sheep farm.

"We inherited Suffolk and Texel crosses, plus Exmoor

Below: Mike and Serena with (L-R) Harley, Isabella and Harry. *(EHFN)*

Ilkerton Ridge Commoners' Meeting
with Natural England and the ENPA

At the beginning of the year, the Ilkerton Ridge commoners entered into a CS Higher Tier agreement to manage the moorland on Ilkerton Ridge. Through the agreement, things like the stocking levels of cattle, sheep and ponies throughout the year and the control of invasive plants are agreed between NE, the owners of the common and the commoners.

The control of European gorse and *Molinia* are top priorities on this area of moorland, which is a Site of Special Scientific Interest (SSSI). However, control is often complicated and costly, and some management options can have unintended consequences. For instance, Shirley Blaylock (ENPA's Historic Environment Conservation Officer) was concerned that a proposed method of gorse removal could damage hidden archaeology. Consequently, the Commoners' Association had a site meeting in July with Mike Pearce (Lead Adviser for agri-environmental agreements at NE) and Shirley to agree the best way forward.

Below: Mike Pearce from Natural England discussing gorse management. *(VE)*

Moorland Event
Organised by FWAG SW and the EHFN

Above: Robin May explaining how Aclands is managed. *(SE)*

Another example of collaboration between farmers and organisations concerned with the management of Exmoor's moorlands was an event organised as part of the Headwaters of the Exe programme by FWAG SW and the EHFN for anyone interested in a practical workshop on the surveys required as part of the SFI Action for Moorlands.

Everyone met at Bray Common, Aclands (which is owned by May Brothers and lies west of Simonsbath), for a practical demonstration of how to undertake a survey using the Foundation for Common Land's moorland surveying app downloaded to a mobile phone. The requirements of the scheme, together with its purpose and ambitions, were discussed before everyone walked out onto the moorland to do some surveying, led by Anne May from FWAG SW.

They also viewed an area of moorland that had been rewetted about 16 years ago through the Exmoor Mires Partnership, which undertook the surveying and restoration work. Robin May was on hand to talk about the work and how the moorland has changed as a result.

The event was organised and funded by the South West Water's Upstream Thinking Headwaters of the Exe project.

August

August is the month for summer shows, and Exmoor's three local breeds of livestock (Devon cattle, Exmoor Horn sheep and Devon Closewool sheep) are all shown enthusiastically.

Angela and Ian Poad's Devon cattle have been doing particularly well in the show ring.

However, TB can be a real problem, both for farmers wanting to sell suckled calves, like Mark Williams, and farmers wanting to show their cattle. If a herd is under TB restrictions, a whole showing season can be lost. This was the case for the Larkbarrow herd from Wellshead, managed by Ricky Atkins. He and his family still had fun exhibiting their Exmoor Horn sheep, though. Showing Exmoor Horns is also a family activity enjoyed by Ross Crang, Lexy Floyd and their two young girls.

Keith Branfield was the Exmoor Horn judge at Dunster Show this year.

Margaret Elliot is well known for her pedigree Devon cattle and Devon Closewool sheep. She has been successful at several shows this year with her Closewools, winning coveted special rosettes to mark the breed society's centenary.

This milestone event was celebrated with a wonderful lunch and farm open day at West Whitefield, Challacombe.

Mark Williams and Family
Lower Blackland Farm, Withypool

Mark Williams is 30 years old and the fifth generation of his family to farm at Lower Blackland, near Withypool.

"Mum and Dad have been really good at letting me make decisions," he says. "When I came back from Hartpury College, I worked part time for a local agricultural contractor near Simonsbath as well as working at home, and I still try to do work off the farm when I can. Dad has gradually been giving me more responsibility. He's let me try my own ideas out, like buying in some Welsh Mules, which have been successful."

In all, there are about 40 spring-calving suckler cows (mainly Hereford, Friesian, Limousin and Angus crosses, which are put to a Charolais bull) and 500 ewes (Texel and Suffolk crosses plus an increasing number of Welsh Mules). Lambing takes place at the end of March, and the whole family rallies round, with Mark's sister Katherine taking a 'holiday' from the EHFN so she can help.

Lower Blackland shares common grazing rights on Withypool Common with several other commoners. Mark grazes ewes there during the summer months, which saves the farm's grass and helps moorland management.

Like so many cattle farmers on Exmoor, he rates bovine

Below: Mark with his parents Pat and Charlie.

Left: Mark and one of his Hereford crossbred cows.

TB as one of the most difficult risks to deal with. "We've had six TB breakdowns since Dad took on the farm in 1973, and the last one was about four years ago," he says. "We aim to produce calves for the two-day autumn Exmoor Suckled Calf Rearers' Association [ESCRA] sale at Cutcombe, so they're weaned and sold at about eight months old. Because our whole system is geared towards that, if we fail our TB test it's really serious. If all those calves have to be kept on-farm until the herd goes clear, that puts a huge strain on the farm's finances and also on shed space and forage supplies during the winter. The whole thing is stressful: not only the worry of failing the test but also the testing itself. We tend to have a pre-movement test, instead of a free whole herd test, as it cuts the chances of us going down because we're only testing the animals we want to sell."

Mark's mother Pat keeps all the old catalogues from markets they've attended, and they provide a fascinating record of how entries and prices have changed. For instance, in 2000, there were 2,500 suckled calves entered in the two-day ESCRA sale, but last year there were just 1,122 calves. The decline in the number of entries may be due to several things, but TB has undoubtedly played a large part.

Mark finds marketing a big responsibility. "You've got to plan so far in advance because of TB testing and other rules and regulations, and you're going into the unknown when you take your livestock to market. You have to take what you get on the day, really, because it's too complicated to bring the animals back home with regards to the six-day standstill."

Mark says he's optimistic about the future of farming on Exmoor as, at the end of the day, people have to eat. Like his father, he takes great pride in producing quality livestock, but he thinks he may have to diversify in some way now that agricultural support schemes are changing. "Trying to get the right balance between the environment and food production is important for the future of Britain's farms," he comments. "I think it's going to get a lot tougher on Exmoor, with increased legislation plus all the uncertainty, but I wouldn't change my farming life for anything."

Left: Charolais bull.

Bridget Goscomb
Farm Vet

Bridget Goscomb moved to Sindercombe Farm, Twitchen, in 1987 with her husband John and six-month-old baby son. Since graduating in 1980, she had been working as a vet in Gloucestershire, where bovine tuberculosis (TB) was first discovered in badgers in 1971, and TB duties were part of general practice.

Less than a year after moving to Exmoor, Bridget was offered a job as a temporary Veterinary Inspector for the State Veterinary Service. Her work involved TB testing as well as dealing with cases of bovine spongiform encephalopathy (BSE).

"When I first started the job in 1988, there were four premises under TB restrictions in Devon," she remembers. "There were hotspots in Hartland and East and West Putford, but nothing on Exmoor." Studies and trials have shown that although many different mammals can become infected with TB, badgers are particularly significant wildlife reservoirs, Bridget explains. They can spread *M.bovis* bacteria via direct transmission as well as through their excreta. Deer can also get TB, but they tend to be end hosts; if they die from TB and mammals like badgers and foxes eat the carcass, that can be a route to infection. The *M.bovis* bacterium is particularly tricky to detect and control because it can form cysts that lie dormant for years inside a host animal.

TB had almost been eradicated from cattle in the UK by the end of the 1970s, through the gassing of infected badgers in their setts in areas where the disease was prevalent, together with the testing and slaughter of infected cattle. However, after the Zuckerman report of 1980, the badger culling strategy changed, and the number of cases began to increase. In 1997, the Krebs report concluded that the evidence linking badgers to the spread of bovine TB was compelling. In spite of this, however, culling was abandoned in favour of a nine-year randomised trial. Bridget, together with many other vets and scientists, thought that the Krebs trial was conducted inefficiently, and she believes that the change in policy, together with the far-reaching effects of the 2001 foot and mouth epidemic led to a dramatic escalation in the incidence of TB. "No testing during the foot and mouth crisis and no pre-movement testing for cattle used to restock farms in the other parts of the country, particularly Cumbria, led to a perfect storm."

Right: Cattle are restrained in a cattle crush while two injection sites are prepared, skin thickness is measured at each site, and avian tuberculin is injected into the skin of the upper site while bovine tuberculin is injected into the lower one. Here, Bridget is measuring skin thickness using callipers.

Since then, whole-herd testing and the pre-movement testing of cattle, together with badger culls in areas with a high prevalence of TB, have made a difference. Exmoor is within a TB 'high risk' area, so whole-herd tests are required every six months and pre-movement tests (over 60 days after a clear whole-herd test) are also required. In 2022, the number of cattle slaughtered in England due to bovine TB was 22,084 – the lowest level for 15 years and a 20% decrease on the previous year. "The system we have at the moment is far from perfect, but it is helping to reduce cases. There are now far more routine whole-herd and pre-movement tests than short interval tests of herds under TB restrictions," Bridget says.

Testing badgers for TB so that only diseased ones are culled is now possible, and she would like the government to take that forward. "Vaccination has a long way to go yet," she adds. "It has little effect on infection rates and is impractical in this area."

Since 2008, Bridget has been working for a local practice in mid-Devon, mainly TB testing.

"It's a huge responsibility. I can shut a business down with one inconclusive test, but I do the test to the best of my ability," she says. "I fully appreciate the stress that farmers go through. We've had reactors at home, so I know what it's like, and I try to offer support and advice when needed. I've been able to get to know many farmers in the area, and I have made some tremendous friendships through testing."

Above: Bulls can be difficult to test. Fortunately, this one was being cooperative.

Left: After testing has been completed, the cattle crush gate is opened and the cow is free.

Ricky Atkins
Farm Manager, The Wellshead Estate, Exford

Ricky Atkins spent some of his childhood at Wellshead, near Exford, when his father Terry was stockman for Mike and Sue Lanz. After that, Terry became a self-employed painter and decorator while also doing lambing work for several people and showing Exmoor Horn sheep for Richard and Elizabeth Crabb at Riscombe. When the Crabbs decided to sell their sheep, Terry bought them and carried on showing.

"Besides helping Dad with his sheep, I spent my weekends and school holidays at Coombe Farm, where Lionel and Mike Stanbury [Serena Colwill's grandfather and father] were very generous with their knowledge and patient with my mistakes," Ricky remembers. "John Tucker at Stetfold Rocks was good to me later on, too; I worked there for a couple of years."

In a way it felt like coming home for Ricky when he became the stockman at Wellshead in 2013. "It was the biggest decision of my life," he says. "It meant giving up the business I had with my brother, shearing in the summer and hedging in the winter, and it also meant giving up working for Tony and John Richards at Yarner plus Roger Webber at Dunkery View, which I loved. A lot of what I know about farming now I owe to them."

Below: Emily and Ricky showing Exmoor Horn ram hoggets (yearlings) at Dunster Show. Ricky has taught his daughters to keep their eyes on the judge when they are showing.

Above: The Atkins family. L-R: Sarah, Imogen, Emily and Ricky.

At Ricky's job interview, the agent for Henry Rawson (the new owner of Wellshead) said he wanted someone who would stop and fix a problem when he noticed it rather than ignoring it and walking past.

"I've always been that sort of person, but I often remember his remark in my everyday work, especially when I see something that needs fixing!" Ricky comments.

When he moved in with his wife Sarah and young daughter Emily, there were just over 40 pedigree Devon cattle from the original herd (called the Wellshead herd) that had belonged to Mike and Sue Lanz.

"Henry wanted to stay traditional, with Devon cattle and Exmoor Horn sheep, which I was very happy about as they're what I grew up with and they suit the conditions here. We've now got about four-hundred-and-fifty Exmoor Horn ewes and a hundred pedigree Devon cows plus followers. They've just got better and better. Apart from the herd being in and out of TB restrictions, we haven't looked back."

The new herd at Wellshead is called the Larkbarrow herd.

Showing is something Ricky grew up with, and he loves it. His enthusiasm is being passed on to his daughters, eleven-year-old Emily and eight-year-old Imogen.

"The showing community is like an extended family," he says. "It's good fun and the kids absolutely love it. It teaches them so much; it's lovely if you win something but not a problem if you don't. We're all just happy for each other."

Ricky shows his own sheep and Henry Rawson's Devon cattle. His involvement with showing doesn't stop there, though, as he judges both sheep and cattle at local and regional shows. He has also taken on organising the sheep section of Exford Show. "I'm doing it in memory of Dad and his friend Robin Hayes as they were such a big part of it. So many more people are showing both Exmoor Horns

Above: Ricky's dog Fern waiting patiently for him on the quad bike.

and Closewools now. It's really good, and the standard is getting better all the time, but it does mean it's becoming harder to do well."

Unfortunately, the Larkbarrow herd has been under TB restrictions until very recently. Although the Atkins' kitchen sideboard is decorated with cups, trophies and rosettes from the 2023 showing season, the only prizes for cattle are the prestigious Devon Cattle Breeders' Society trophy for the best large herd of pedigree Devons and a smaller trophy for the herdsman of the winning herd. Judging for the herd competition takes place on-farm, so it's something any breeder of Devon cattle can enter, even if their herd is under TB restrictions.

The Larkbarrow herd's award indicates how well the cattle could have done this year if they had been allowed to travel to shows. They passed their final short interval test two weeks before Dunster Show, but by then it was too late. A whole season of showing had been lost, along with some potential sales of top-class pedigree breeding stock.

When asked if he has a favourite trophy from the vast array on the sideboard, Ricky picks a cup without hesitation. "We gave this in memory of my dad and his great love for the Exford Flower Show. It's the Terry Atkins Memorial Cup for the overall best in show entry," he explains. "I baked a cake in honour of the King's Coronation for the Men Only Baking class, and then went on to win this. It was a great honour. I know Dad would have been so proud."

Above: Ricky checking the cattle on their summer grazing at the top of Wellshead.

Above: A fine flock of Exmoor Horn sheep at Wellshead.

84

A calf in the heather on Winsford Hill.

Angela and Ian Poad
Furzemoor Farm, Exford

Ian and Angela Poad from Furzemoor Farm started their Exmoor herd of pedigree Devon cattle 10 years ago, and already they've got one of the most highly regarded herds in the country. It would be a mistake, though, to put that down to beginner's luck. A wealth of experience breeding, rearing and showing cattle has led to their success. Angela's father was the renowned Devon cattle breeder Andrew Slee, owner of the Halsbury herd, and Ian used to be a dairy farmer in Cornwall.

The Poads bought Furzemoor with eight acres in 1998, and in 2008 they began showing cattle from the Wellshead herd for Mike and Sue Lanz – with considerable success. Then, just after Wellshead was sold, 35 acres of land next to Furzemoor came up for sale in 2013, so Ian and Angela bought it and wasted no time in buying some Devon cattle.

They now have 18 cows plus followers. The majority of heifers they breed are kept as herd replacements or sold for breeding, while the best males are kept as bulls and the rest are sold via an agent to specialist butchers.

"It takes time to start a herd from scratch," Angela says. "We were always going to have a small herd, so we've concentrated on quality rather than quantity. We want the type that's on the larger side of medium in size, with good length, bone, breed character and the ability to convert feed well, too."

Below: Ian and Exmoor Captain waiting for their class.

Below: Angela giving Exmoor Captain's tail a final comb before they enter the ring at Dunster Show.

Above: "There's an art to showing." Angela and Ian showing Exmoor Captain at Dunster.

"The important side of any breeding is the female side," Ian continues. "When buying a bull, we like to see the mother and, if possible, the grandmother as well. Our top bull Captain's grandmother lived until she was twenty years old and never needed her feet trimmed."

"Feet, legs and udder, including teat placement and size, are all so important," Angela agrees.

"And temperament," Ian says. "That's partly genetics, but also how they're handled."

The Poads spend as much time as possible with their cattle, and they make a point of walking amongst them in the field rather than just driving around to check them.

"To an outsider, showing looks easy, especially when it's done well, but it takes a lot of time and dedication," says Angela, who learned a lot from her father and helped him show cattle all over the country, including at the Royal Smithfield Show when it was at Earl's Court in London.

"There's an art to showing," Ian remarks. "You may be able to lead cattle, but showing's a bit different. Some cattle like the whole process better than others – the good ones take your eye, but others just go through the motions. You've got to get inside the animal's head; you and the animal have got to become one."

Angela nods in agreement. "They have to trust you. You're taking them away from everything they know at home to a show where everything's strange. You can prepare them as much as possible – getting them used to loud noises and flappy things – but the whole show environment is scary. Trust is essential."

She wonders whether the amount of time it takes to prepare cattle physically and mentally for the show ring may be one of the reasons why cattle classes struggle to get a lot of entries nowadays. Also, Covid has probably played a part because two years without any shows may have got people out of the habit of showing, as well as breaking the cycle of training calves and giving them experience from an early age. Then there's the cost-of-living crisis and last, but not least, there's bovine TB, which has had a huge impact.

For the first time this year, Angela and Ian have shown only home-bred cattle, and they have had their best results ever. Captain was Breed Champion seven times and his son Forrester was Breed Champion once. In addition, Captain was Supreme Champion at the Mid Devon Show and Reserve Supreme Champion at Dunster. Furthermore, their cow Exmoor Flirt was Female Champion at both Devon County Show and the Bath and West Show.

"We've made a lot of friends and met a lot of interesting people – other cattle breeders as well as Devon breeders," Ian says. "And it's a great way to promote your herd as well as the breed. This year, we sold four bulls as a direct result of people seeing Captain at shows and wanting his sons. Going to shows keeps our herd fresh in people's minds, especially if there's a prize-winner."

"We love our cattle, and we really enjoy showing them," Angela concludes. "If we didn't enjoy showing, it would just be an awful lot of hard work."

Later in 2023, Captain became the Devon Cattle Breeders' Society Bull of the Year, Forrester was Young Bull of the Year and Flirt was Female of the Year. The Poads also won Breeder of the Year.

Below: The Exmoor herd relaxing at home. Their fields have tremendous views over Exford. *(VE)*

Ross Crang and Lexy Floyd
Chiltons Farm, Dulverton

Ross Crang and Lexy Floyd have been renting Chiltons Farm for three years. "We were very lucky to get the tenancy, Ross says. "When we first got together in 2017, we rented John Nott's house in Stoke Rivers and various bits of land all around the countryside. It makes a huge difference having everything in one place here."

Ross grew up at East Challacombe Farm, Combe Martin, with his parents and sister. His family ran a herd of Charolais cattle and a flock of breeding ewes. "Farming is all I've ever wanted to do, but the home farm was never quite big enough," he says.

Lexy spent her early childhood on Exmoor before moving to London with her mum for her school years. However, she often came back to Exmoor for the holidays, staying with her uncle, aunt and cousins at Hallslake, a farm near Brendon.

"From an early age I wanted to live on Exmoor and do farming, so after GCSEs I moved down here," she recalls.

She worked for a couple of local farmers, farming as well as riding their hunters and point-to-pointers, before working on farms in Australia and New Zealand for a couple of years.

Below: Ross with Daisy and Lexy with Ella

Margaret Elliott
Oxgrove Farm, Brompton Regis

Margaret Elliott (née Luxton) was born into farming. Her parents lived at Oatway Farm, where they kept Devon cattle and Closewool sheep, and she has remained loyal to both of these breeds ever since.

In 1966, Margaret married Donald Elliott, a vet. As a wedding present, her grandfather gave them four fields and a derelict cottage that had been uninhabited for 25 years and had no utilities. It was 600 yards from the road, with no vehicular access. Later, in 1970, they bought more land from Margaret's mother. They called their farm Oxgrove.

Donald worked in the Dulverton Veterinary Practice for forty years and, to begin with, Margaret had to answer phone calls from clients on the phone at home. "It wasn't easy," she remembers. "I had two young daughters and there was a bell on the outside of the house so I could hear the telephone ringing if I was outdoors. There are eighteen steps down to the house, so answering calls kept me fit!"

Before 2001, Margaret and Donald had increased their livestock numbers to 50 cows and 300 ewes, with extra grass keep away from the farm. However, when there was a ban on livestock movements during the foot and mouth crisis some of their animals became stranded away from home, which was incredibly stressful. After that, they sold a lot of cattle and cut their livestock numbers back to what they could keep comfortably at Oxgrove Farm.

"Prior to foot and mouth, we used to sell quite a lot of bulls, and some did very well," Margaret says. "We kept my father's herd prefix of Cutcombe and bred true to the original type of Devon, without introducing polled or Salers genetics."

Devon cattle from the Cutcombe herd feature in many top pedigrees, both at home and abroad, and this year Cutcombe Lucky 24th was the top-priced female at the Spring Show and Sale of Devons at Sedgemoor.

Margaret now has around 10 pedigree Devon cows plus heifers, and 50 Closewool ewes plus two-tooth ewes.

"There's no money in farming – not with the amount of acres we've got – but It's a hard thing just to give the whole lot up," she says. "I've made a lot of friends in both the Devon Cattle Breeders' Society and the Devon Closewool Sheep Breeders' Society. My daughters are both busy doing other things, but one daughter comes around to help. Also, my brother Edward Luxton helps me with showing."

Above: The Oxgrove Flock display board at Dunster Show. (The Closewool Centenary herd display boards were funded through FiPL.) *(Melanie Davies)*

Margaret has always enjoyed showing, but there has never been enough time to go to the big shows. Now that she's in her seventies, a lifetime of breeding Closewools is being rewarded with tremendous successes at local shows and top prices in the sale ring.

This year, she won Breed Champion at the Mid Devon Show with a shearling ram, and Breed Champion with him at Dunster Show as well. Her six-tooth ewe won the Old Ewe class at Dunster and she also won the class for the Group of Three.

"Dunster has always been a good show for Closewools; if you win something there, you haven't done too badly," she says with typical modesty.

At Exford Show, the week before, Margaret's sheep also did exceptionally well. Again, her shearling ram was first

Above: Margaret taking part in a parade of prize-winning livestock in the main ring at Exford Show. *(VE)*

in his class as well as Male Champion, and there were also firsts in the Old Ewe and Group of Three classes, plus second prizes for another shearling ram, her old ram and her ram lamb. Her six-tooth ewe then went on to win Female Champion and Interbreed Champion.

"That ewe has done me so well. I only picked her out a few weeks before, when I weaned the lambs. She's had twins both lambings, so she's worked for her living," Margaret says. "It's wonderful, really – a lovely memory to have."

Right: Margaret receiving a trophy from Emily Atkins at Exford Show. *(VE)*

Keith Branfield,
Westwater Farm, Withypool
A Livestock Judge's Point of View

"Being a livestock judge is definitely a bit nerve-wracking beforehand, but once the class starts and the competitors have appeared in the ring, it all begins to fall into place," Keith says. "Judging the Exmoor Horns at Dunster this year was especially difficult because there were so many entries and the standard was very high. It's terrific how many entries there were – a spectacle to behold. My time was consumed by trying to pick out the top seven or eight from classes of twenty plus, but I really enjoyed it. The sheep were a credit to all the people who put in the effort to enter, and it was great to see so many fresh faces as well as old friends. It's especially good to see so many children showing – we've got to encourage the young ones."

From the age of twelve, Keith lived with his mother, father and brothers on a livestock farm near Hawkridge. As a teenager, he joined Young Farmers and learned many skills, including livestock judging. In time, he was picked for a YFC team that went around major shows competing in judging competitions. "We had to justify why we'd placed the sheep or cattle before us in that order, and we were scored on our decisions, which taught me a tremendous amount," he says. "Those competitions give a very good grounding for future livestock judges."

In 1982, Keith married Sue Clatworthy. They moved into Westwater Farm, near Withypool, taking on a small part of the farm and 60 sheep – half of which were Exmoor Horns. Sue provided bed and breakfast for visitors in the farmhouse and Keith went shearing, and then they rented more land from Sue's parents.

Tragically, Sue died in 2010. Their two children, Robert and Julie, are married with families of their own now, and both of them live away from Exmoor.

"I'm farming alone, so I've sold the cattle and cut back to about a hundred-and-eighty Exmoor Horn ewes. Half of them are put to Exmoor Horn rams and the other half to Bluefaced Leicesters," Keith explains. "I suppose it would be sensible to give up, but farming's a total way of life and I love living here. If I sold up I'd never find anywhere better,

Above: Keith at Dunster Show. *(Melanie Davies)*

and I'd like to keep this place for the grandchildren. It's sometimes hard having to do everything myself and also not having anyone to discuss things with, although I know I can always ring up Robert and Julie."

The Exmoor Horn Sheep Breeders' Society is very important to him, and he has made many friends through it. "You know what John F Kennedy said, 'Ask not what your country can do for you – ask what you can do for your country'? Well, that's what I feel about Exmoor. I've always tried to be pro-active, helping out wherever I can. My father-in-law was like that. Exmoor Horn breeders look out for one another, like a family. We've all got something in common even though we lead different lives.

Above: Keith judging Exmoor Horn rams at Dunster Show. *(VE)*

Above: Edwyn Mills being congratulated on his prize.

Above: "It's especially good to see so many children showing – we've got to encourage the young ones."

I've held various posts over the years, but I'm really proud to be President this year. I wasn't expecting it – not yet, anyway – I didn't realise I'd got so old!"

Although Keith knows most of the people who show Exmoor Horns, he never looks at the handler when he's judging sheep. "It's tough sometimes, especially if there are young children in the ring, but everyone's competing on equal terms."

He tries to accept every invitation to judge livestock, whether at a local show or a major one like the Bath and West, Royal Welsh, Devon County or Royal Cornwall.

It's interesting that the attributes he looks for when buying rams for his ewes are often different from those he would prioritise in the show ring. "When I'm buying for myself, I'm looking to complement or correct the traits of the whole flock," he explains. "I bought a champion ram one year and he did my flock a lot of good, but the daughters of another had peculiar Horns, even though he looked fine. A ram can either make your flock or spoil it."

95

Devon Closewool Centenary Event
West Whitefield Farm, Challacombe

In the late 1800s, farmers found that crossing a Devon Longwool with an Exmoor Horn produced a robust sheep with a dense fleece, valuable for both meat and wool. By the early-twentieth century, these sheep had become known as Devon Closewools. They were so popular that they were recognised as a breed in their own right, and The Devon Closewool Sheep Breeders' Society (DCSBS) was formed in 1923.

Plans for the centenary in 2023 started about a year ago, coordinated by the Society's Secretary Kim Dart. She researched and collected a huge quantity of photos and historical information from the past 100 years and, with Jean Kingdon and Pat Burge, put it all together in a series of display boards. Special centenary rosettes were produced for the summer shows, and the Kingdon family offered to host a centenary celebration at West Whitefield Farm, near Challacombe.

Sadly, Ray Kingdon died on 31st May 2023, but his wife Jean and children (Janet Bale, Nigel Kingdon and Glenda Kingdon) decided to carry on with the event at the farm.

So, on Sunday 20th August, over 200 DCSBS members and guests converged on West Whitefield for a wonderful get-together. The event was opened by Lord Clinton,

Above: L-R: Janet Bale, Hannah Bale, Nigel Kingdon, Jean Kingdon, Steve Bale, Glenda Kingdon and Daniel Bale. *(VE)*

whose great-grandfather was the first President of the Society, and a delicious lunch was provided by Linda Pugsley and her team of helpers.

"It was all done for Dad," Janet said afterwards. "We were a bit lost in June, and the weather was so hot, but Dad had

Below: Lord Clinton opening the event. *(VE)*

Below: Richard Clark demonstrating how to trim a ram for the showring. *(VE)*

Above: Peter Huntley conducting an auction for a ewe lamb bred by Ray Kingdon, which was bought by Martin Tucker. *(VE)*

Above: Tim Nicholas from Marwood giving an impressive sheepdog demonstration. *(VE)*

told me what he wanted to do, so we tried to carry out his plans as much as possible. He was a perfectionist and would have made everywhere spotless, but we did our best. I hope he'd have been proud."

"We had a lot of help from family and friends, and a working party from the Closewool Society came over, too, which was lovely," Jean added. "Kim did a fantastic job organising everything. She worked hard to find people who were related to the original members, and Jan Witheridge did name badges for everyone in her beautiful handwriting, which helped no end. It was lovely to see old friends, and to meet so many enthusiastic new members. All in all, it was a super day and in the end the weather was kind to us."

Below: A farm walk to see the sheep. *(VE)*

September

There was a heatwave in the first half of September, which set a number of new temperature records throughout the UK before the spell ended with thunderstorms.

For John Richards at Yarner, this is one of the best months for repairing the walls that are a well-known feature of his farm and have become a labour of love.

Steve Coates used to have a hedging and fencing business, but recently he and his wife Rosanne have built up a successful enterprise producing top-quality free-range eggs.

On a much smaller scale, Fran Bullard has found a niche market for rare breeds of poultry.

September is also a time for selling livestock, including the sheep breed society shows and ram sales. This year, the champion Exmoor Horn and Devon Closewool rams came from two different farms near Challacombe, and they both made record prices in the sale ring.

Newly purchased rams will have a few weeks to settle in their new homes before tupping begins.

The Kingdon and Bale Families
West Whitefield Farm, Challacombe

The Kingdon family came to Whitefield Barton in 1961, just before the cold winter of 1962-3. Bill Kingdon brought with him his flock of pedigree Devon Closewool sheep, which were first registered in the Flock Book in 1951 and later taken on by Ray in 1995.

In 1994, the farm was split between Bill's two eldest sons. Ray and his wife Jean built a new bungalow and named their farm West Whitefield while younger brother David stayed in the farmhouse at Whitefield Barton with his wife Rosemary.

"Ray was always keen on breeding and showing the Closewools so, when the farm was split, they were transferred to him," Jean remembers.

Their daughter Janet met Steve Bale in 2001, when she was living and working at Brendon Manor riding stables. Steve's grandfather Ben Bale was a keen Closewool breeder and judge, and his father Fred was also a good stockman who farmed at Western Ball, Sandyway. In 2004, Steve took on his grandfather's flock and he and Janet began to farm at West Whitefield with Ray and Jean in 2012.

"Dad taught us how to breed and show good sheep," Janet says. "In fact, we joked that he taught us too well because last year we were placed in front of him and won the cup for the champion ram at the Closewool Show and Sale!"

A week or so before Ray became seriously ill after a stroke, he and Janet looked at the sheep with a view to the showing season ahead. After he died, Janet and Steve picked out what they considered to be Ray's two best rams and prepared them for showing. In September, one of the rams won the Closewool Show and Sale at Blackmoor Gate and topped the market at 1,100 guineas, which was the highest price ever for a Closewool ram.

"The emotion was unbelievable," Janet says. "I was in floods of tears. Dad would have been so delighted. I couldn't be more pleased that I'm cleaning the cup we won last year so it can be presented to Mum this year for Dad's ram."

Below: Jean and Janet at the Closewool Ram Sale with the champion ram that was bred by Ray. *(Kim Dart)*

The Ridd Family
West Mead, Challacombe

David and Mary Ridd Peter live at West Mead, Challacombe, with their son Peter, where they farm 240 acres. They used to breed both sheep and cattle, but they've only got sheep now because they sold all their cattle after the 2001 foot and mouth epidemic. Nowadays there are 370 sheep all together: 125 pedigree Exmoor Horns and 250 North Country Mules.

"I've been involved with the Exmoor Horn Sheep Breeders' Association Council for a long time, one way and another, and so has Father," Peter says, "I do enjoy showing and marketing our sheep – it's nice to see people. I don't go to the big shows – just the local summer ones: Mid Devon, North Devon, Exford and Dunster. It's a bit of fun."

The Ridds' Exmoor Horns are sponged in September so they lamb in February and the lambs are well-grown for both showing and selling. Their sheep have a very good reputation.

"We don't always win, but we're picked out quite a lot," Peter says modestly.

This year, he won the class for the Best Group of Three at Exford Show with a ram and two ewes. Then, on 22nd September at the Exmoor Horn Sale at Cutcombe, a ram belonging to D&M Ridd & Son made the top price of 1,050 guineas.

Above: Peter and Mary Ridd with their champion ewe at Exford Show. They also won the first prize (a shepherd's crook) for the best group. *(VE)*

"We've never made so much money – it kept going up and up!" Peter exclaims. "David Snell bought him."

All the lambs that aren't kept as flock replacements are sold as stores at Blackmoor Gate Market, which is only a few miles away. This year they have been making just under £100 a head.

Despite the current changes in national farming policy, Peter reckons they'll keep going much the same at West Mead. "If it works, stick with it," he says.

Given his track record at both shows and sales, that seems a wise philosophy.

Left: David Ridd with the Ridds' champion ram at the Exmoor Horn Ram Sale at Cutcombe. L-R: Richard Dart (judge), David Ridd with the Ridds' champion ram, Stuart Routley with his reserve champion ram, and Darren Potter (judge). *(Kim Dart)*

John Richards
Yarner Farm, Culbone

The views from Yarner Farm are spectacular: neat fields dotted with livestock set against a backdrop of Porlock Bay, with wooded combes and moorland beyond. All the elements of Exmoor brought together in a perfect picture.

"I know it's lovely, but the trouble with farming is I'm looking for problems most of the time rather than anything else," John Richards says.

With about 700 acres to look after, 50 spring-calving cows and over 1,000 sheep that lamb from the end of March until the end of April, John has his work cut out, even though he has help.

"Scottish John is amazing – a proper shepherd – a lifesaver – and his son Matt helps with things like night lambing."

John's wife, Jen, helps when she can, but she works in the Malmsmead Gallery and she also has a theatre company called TakeThreeGirls. Her twenty-one-year-old daughter, Ella, lives at Yarner, while Ella's twin brother Theo is a few miles away in Minehead. John's daughter Martha, from his first marriage, stays at Yarner every other weekend and half the school holidays. "She's great, and helps with whatever I'm doing around the farm, but particularly loves

Above: Jen and John, with Raven the dog.
Below: Yarner Farm from Withyclose.

Above: "This is a good example of what happens to a hedge if it's allowed to just manage itself with no maintenance."

me: in the space of a few minutes, you've had a good year or a bad year. So much depends on things beyond your control, like which buyers turn up and where you are in the catalogue. Markets are a valuable cog in the wheel of marketing, though, as they do set a price."

As well as having a reputation for good-quality livestock, John's reputation for walling and hedging is second to none. "It's my way of expressing my artistic nature – my form of art, without a doubt."

Like his grandfather, John leaves a unique mark in his walls. "As a general rule, you put big stones at the base of a wall and progressively smaller ones as you go up, so the weight of the stones on top doesn't force out the ones underneath," he explains. "I like to put a big ugly stone somewhere near the top of every wall I make, though, about three times the height of the stones around it."

The quality of the stones around Yarner isn't great for walling. "You've either got very hard blue stone from around Culbone or lighter brittle stones that aren't a good shape and shatter when you try to alter them, so most of my walling stones have come from a sixty acre field near Exford that my grandfather bought," John says. "I must have hauled back thousands of tonnes of stones from there – good sandstone you can shape well."

He reckons that the relatively poor quality of the stones on Exmoor is probably the reason why walls around here are stone-faced earth banks rather than dry-stone walls. Most local stones wouldn't stay put without soil, but earth banks alone would be worn away by livestock so they need to be faced with stone.

"A good wall will be there for a century at least without having to be repaired," he remarks. "So you'll never know whether you're a good waller because you'll be dead before you can find out."

The best months for repairing walls, as far as John's concerned, are May, August and September, when he's a little less busy. "I prefer walling in the summer because I can't abide doing it when everything's muddy. Part of the pleasure of walling is noticing all the creatures that live there and in the hedge above, from the tiny bugs nobody really thinks about to small mammals and birds – a classic wildlife corridor… It's hard to beat a sunny day, nature all around, a wall to fix and Test Match Special on the radio. Better than any holiday."

Jen doesn't agree. She's just booked a family holiday in Spain at the end of September, and it's non-negotiable!

Opposite: Father and son: Tony and John, with Toby the dog.

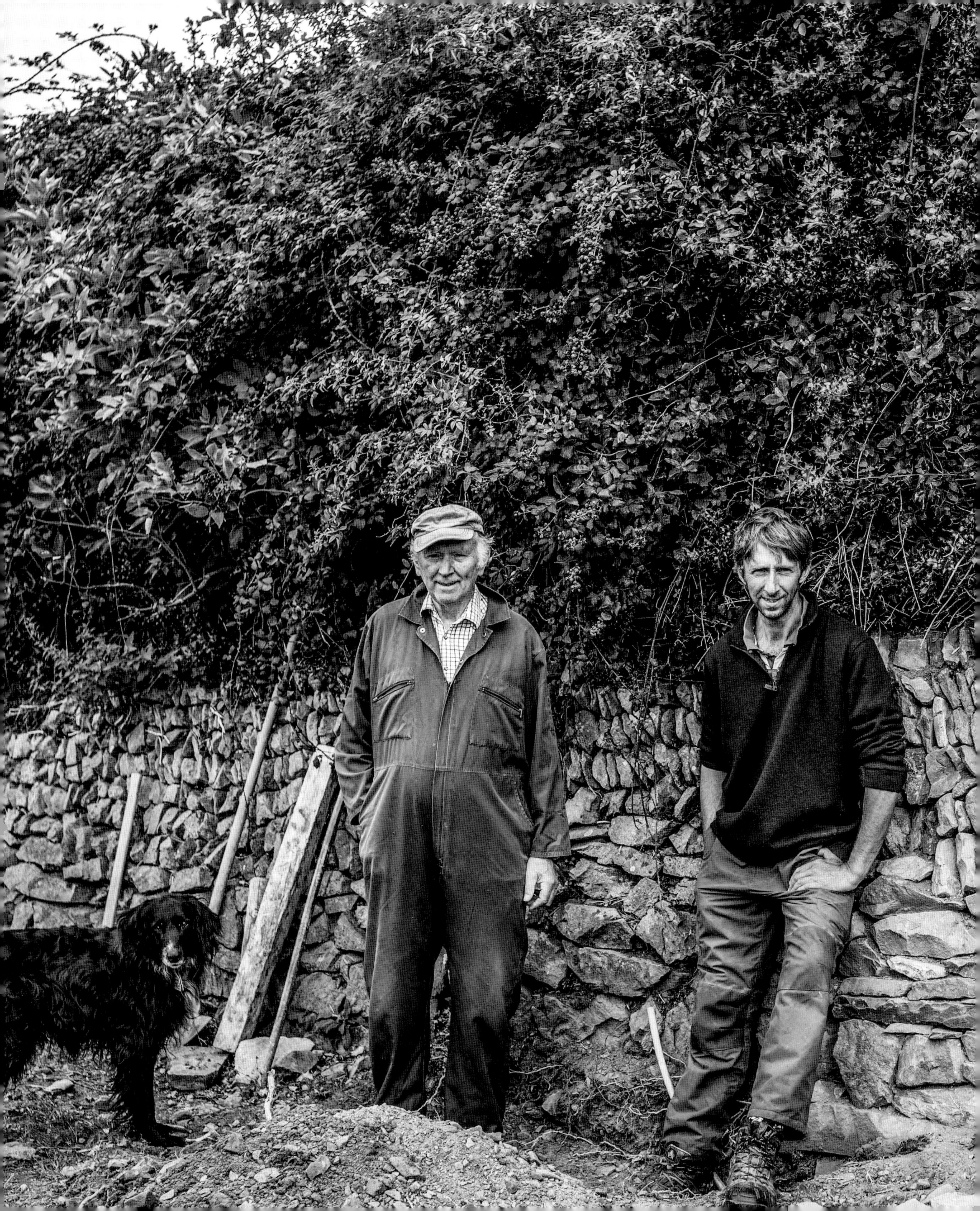

Steve and Rosanne Coates
Woolcotts Barn, Brompton Regis

Rosanne Andrews (as she was then) moved to Lower Woolcotts as a two-year-old when her parents, Bob and Sue, bought the farm and 120 acres. Bob Andrews, a well-known horse dealer, gradually increased the amount of land he owned to 620 acres.

Steve Coates, the eldest son of Anthony and Brenda Coates, was brought up at Chapple Farm, between Bury and Skilgate. Claude Govier, Steve's grandad, became friendly with Bob and taught him about sheep farming. His other grandad, Joseph Sydney Coates, came to Exmoor to be Frank Green's chauffeur and valet at Ashwick, and was awarded the British Empire Medal for brave conduct in the Second World War.

Both Rosanne and Steve remember the reservoir at Wimbleball being built and all the changes that went with it. "It was a beautiful valley with the River Gurney running through, and we got the water for the farm from it," Rosanne says. "A hundred acres were taken from my parents through compulsory purchase. They didn't get much for it, and after the dam was built we had to pay for our water." She remembers going to see the building work every week with Brompton School. The reservoir opened in 1979.

Steve went to boarding school at Brymore and Rosanne went to St Audries in the Quantock Hills. "Letters were our main means of communication with the outside world, something really special," Steve recalls. "Rosanne and I became pen pals – well, Rosanne and three other girls, actually. They all used to write to me, and I wrote back. Some friends and I escaped from school one night, intending to walk to St Audries to see them, but we gave up after a couple of miles!"

They married in 1995, and have been farming from Woolcotts Barn for past 26 years.
Steve had a business doing hedgebank restoration, hedging and fencing, while Rosanne did night lambing from February until April and worked in the tea rooms at Wimbleball Lake, plus took horses on livery. On top of that, they had to juggle looking after their three children Trevor, Charlie and Sarah.

Below: Steve and Rosanne standing on the dam at Wimbleball Lake. They have to cross Bessom Bridge every day to get to land they farm on the other side.

Above: View of the River Haddeo, and woodland belonging to Bernard Dru, from the dam at Wimbleball Lake.

The ESA scheme was launched on Exmoor in 1993. Amongst other things, farmers were paid capital grants for managing their hedges, and that provided a huge amount of work for people like Steve. His hedging and fencing business grew until he was employing four people to help him. However, almost overnight the funding for capital works like hedge restoration dried up just before ESA schemes were replaced by Environmental Stewardship (ES) in 2005, and after that far fewer farmers could access grants for capital works.

"I had to wind up the business and let the lads go," Steve remembers. "We were still farming sheep and cattle but there weren't enough acres for that to pay, so we had to decide whether to walk away or stay and fight. We arranged a meeting with a farm consultant, and he suggested pigs or egg production. I said, 'Not on your nelly!' but when we'd had a bit of time to think about it, we came round to the idea of poultry."

Above: Surplus water from the reservoir flows down a waterfall into the River Haddeo.

They started their planning application for two poultry barns in 2015, and got their first consignment of 8,000 birds in June 2017 after a lengthy planning process.

Rosanne takes up the story: "In December 2017, there was an awful storm with eighty-mile-an-hour winds that swept up the valley. We lost one shed and half the birds in it. We put the survivors in the other shed, but that meant there were too many in there so we had to cull quite a few. It was terrible. I wanted to give up."

"Carrying on with just one shed wasn't really an option; without two sheds we couldn't have two lots of birds of different ages to even out the cash flow and we couldn't keep enough birds to make the whole enterprise worthwhile," Steve explains. "So we were faced with either selling up or trying to get a new shed. We wanted to keep the overall footprint of the ruined shed but change its orientation, to minimise the risk of the same thing happening again, but that required full planning permission, which took nine months and cost a fortune in fees and consultancy bills. Also, the cost of the shed went up by fifty per cent in the time it took to get permission."

Above: Hens are free to roam outside in a large, well-fenced field during the day. Trees have been planted to provide shelter.

In the midst of all that, the unthinkable happened: Rosanne and Steve's eldest son, Trevor, died in December 2020 after the car in which he was a passenger crashed. He was 23 years old, and he and his partner Chloe were expecting a baby. Ivy Rose Coates was born on 8th January 2021.

Steve says it's good to talk about Trevor to keep his memory alive. "It nearly finished us. Everything was put on hold for a while. We've had to be very strong in our minds to keep going," he says. "David Weir was fantastic, and so was the vicar of Dulverton, Dowell Conning."

Less than a year later, another local young man who was a passenger in a car was also killed. In response, David Weir and Katherine Williams, together with Steve and Rosanne, organised a road safety event at Cutcombe Market, which was attended by the emergency services and people from the Exmoor area.

"Speaking at that event is the most powerful thing we've ever done," Steve says. "There were many more people than we'd anticipated, and their attention was focused on us in the auction ring. Making myself reflect on what had happened choked me, and I wasn't sure whether I'd be able to put everything into words, but I'm very glad I did it. If it helps to save one life it will have been worth it."

Rosanne agrees. "Yes, we've all got to look after each other. I'd like to think that some sort of taxi service could be started up on Exmoor, so nobody's tempted to drive home after they've had a drink; perhaps some sort of buddy system, which is what Dowell Conning suggested."

For most of his life Steve has been involved with hunting in some way.

Left: Burford Brown hens are calm and non-aggressive.

Above: Inside the hen house, where the flock goes at night to keep safe. The hens lay their eggs in nest boxes in the hen house.

Left: Eggs roll from the nest boxes onto a conveyor belt that takes them away so they can be collected and packed into trays.

Below: When a tray has been filled, the eggs are stamped with the farm's unique number for traceability. The trays of eggs are then put on a pallet in stacks of six, ready for collection.

"When we lost Trevor, Rosanne knew I had to find something else to focus on – an escape," Steve says. "I was asked to become a joint Master of the Devon and Somerset Staghounds, and I accepted. This will be my third season, and I'm thoroughly enjoying it. I do one or two days a week. And then there's cricket in the summer, with a completely different lot of friends." He smiles at Rosanne. "We've just got to find something to take you away from the farm now."

She replies that there's so much to do at home, and she still finds going out very difficult.

Steve and Rosanne have 200 Texel cross ewes and 40 suckler cows to look after as well as their poultry enterprise. They admit that if they'd known beforehand how much paperwork would be involved, they probably wouldn't have embarked on the egg project. "To start with it was a bit daunting, but it's grown on us and we get very attached to the hens in our care," Steve says. "It's not just about collecting and packing eggs – it's mostly about animal husbandry, so it's similar to cattle and sheep farming in many ways."

At the moment there are about 10,000 birds in all: 6,000 birds that have been laying for several months in the new shed and 4,180 point-of-lay-pullets in the older one. The two flocks are completely separate, and there's strict biosecurity between the two sheds.

Their contract is with Stonegate, owner of the brand Clarence Court. Stonegate has developed the Burford Brown chicken, which is genetically adapted to be calm and non-aggressive towards others in the flock. Meticulous record keeping and attention to detail are needed to meet the high welfare and quality standards required.

"The layers are most expensive when they first come in at fifteen to sixteen weeks old – they have to be kept inside for a while, and they eat without producing any eggs until they're nineteen to twenty weeks old," Steve explains. "Then, for the next three weeks or so you get small eggs that aren't worth much, so there's no income to speak of for about two months. After that, the hens carry on laying until they're seventy-six weeks old."

Both Rosanne and Steve agree that the biggest risk to their business is bird flu. "It would wipe us out if our hens got it – we'd lose our flock and our income for a whole year," says Steve. "Our biosecurity is as tight as possible, but birds from the nearby shoots are a massive threat, and once they're released they're classified as wild birds. One of the release pens is only five hundred metres from our poultry units. It's certainly a challenge, but everything in life nowadays is a challenge, isn't it?"

Fran Bullard
Fran's Free-Range Poultry

Withypool is where Fran grew up and it's where she lives now, running her free-range poultry business from The Firs Bungalow, a smallholding that belongs to her dad and mum, Steve and Diane. She breeds and sells an amazing array of poultry, and also sells eggs and homemade cakes in her parents' roadside shop.

Six years ago, Fran met Vinny Webster at the Great Bradley shoot. He was a jockey from Ireland who was working for Withypool-based trainer Nicky Martin, and she was working for Richard Smaldon at Exmoor Agriculture, near North Molton. Nineteen months ago, they had a baby girl called Willow.

Vinny has just given up working with horses full time, although he'll still ride at weekends, and he's re-training as a bricklayer. At the moment they're living on the site of the business in Withypool, but they'd love to build their own house and are working hard to turn that dream into reality.

"I wanted to do something where Willow could come too, so I began doing cleaning jobs, including cleaning for Keith Branfield, who got me back into lambing again. He's lovely to work for. I bake him cakes and he gives me his homemade marmalade," Fran says. "And alongside my cleaning jobs, I started breeding poultry."

To begin with she just had Ayam Cemani chickens (an

Below: Willow holding an Ayam Cemani chicken.

Above: Fran and her daughter Willow, with Mini the Cocker Spaniel.

Indonesian breed that's almost completely black – even through to the meat and bones), Friesian Fowl (a light breed that's good at foraging and isn't too broody) and Ixworth chickens (a white, fast-growing meat breed with good laying qualities). Then, as demand increased, she purchased a breeding group of Blue Partridge Brahmas from championship bloodlines as well as Gold Laced Wyandotte bantams, Silver and Gold Laced Sebrights, and several other breeds. In addition, there are Call ducks, Runner ducks, Khaki Campbells and quail.

"It's been a job keeping up with it all. I can't quite believe how well it's gone. Customers have even come from London. People are so lovely – they keep in touch and send photos. I also sell eggs on eBay and send them by next-day delivery, which works surprisingly well, and I'm getting a lot of local customers now who are really glad they haven't got to travel miles to buy poultry anymore."

Above: A waddling of Call ducks.

Above: These Gold Laced Wyandottes and Lemon Millefleur Sablepoots are about 12 weeks old.

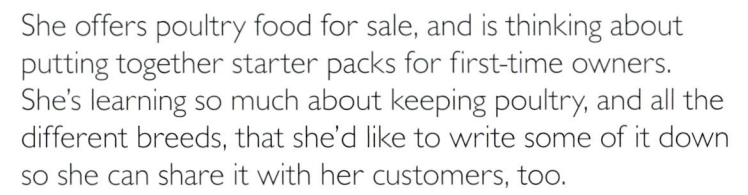

Above: Runner ducks.

She offers poultry food for sale, and is thinking about putting together starter packs for first-time owners. She's learning so much about keeping poultry, and all the different breeds, that she'd like to write some of it down so she can share it with her customers, too.

Fran's pens and coops are homemade from pallet wood and crates, to recycle as much as possible and save money, but although they're designed to be vermin-proof, predation by foxes can still be a problem. Another big threat to the business is bird flu. She's geared up for the inevitability of 'flockdown' regulations again this winter, but it makes keeping poultry (especially ducks) much more difficult, and the possibility of the birds getting flu is a constant worry.

"Despite the problems, I'm so glad I started my poultry enterprise," Fran says. "Willow adores the birds. She's so excited to come out and help me every morning, and that makes it all especially worthwhile."

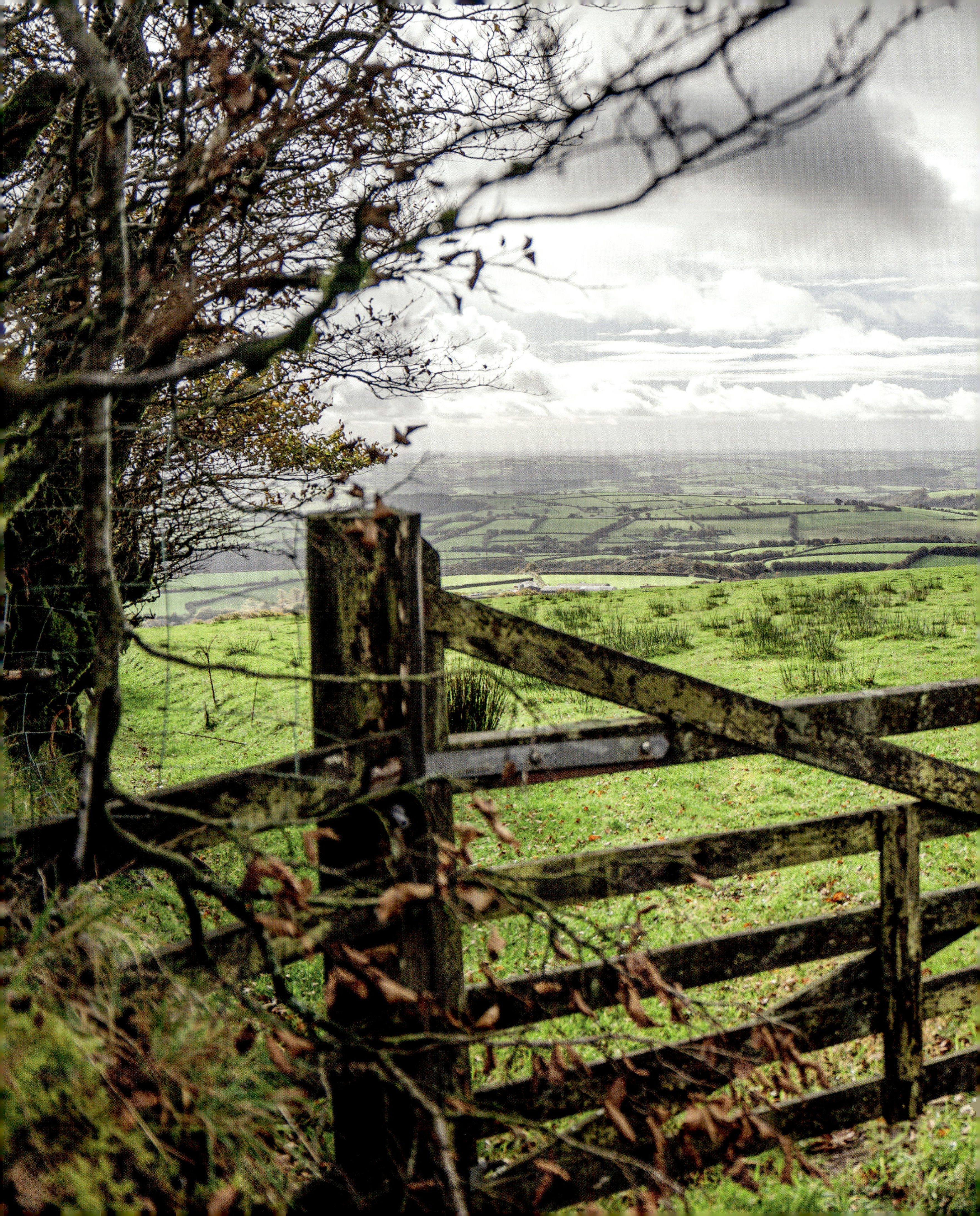

View south from North Molton Ridge.

October

The harvest is celebrated all over the UK in October. For the majority of farmers on Exmoor, this means selling the year's crop of sheep and cattle, so this is a busy month for Peter Huntley and the team at Exmoor Farmers Livestock Auctions (EFLA).

Traditionally, livestock bred on Exmoor are sold in the autumn to farmers from lowland areas. The two-day sale of suckled calves at Cutcombe is an especially important fixture for many, including Jeremy and Angela Andrews.

Markets are often where sheep scab is accidentally transmitted between flocks. The most effective way to deal with scab is by dipping, and Julian Branfield supplies a useful mobile dipping service.

October is also when free-living herds of ponies are rounded up for the foals to be weaned, inspected and (if all is well) registered as pedigree Exmoor ponies.

The Woolhanger Sheepdog Trials, organised by the famous shepherd David Kennard, is an unmissable event for anyone interested in working sheepdogs. Puppies with good working parents generally make the best sheepdogs, and Rabbit Slatterly has bred and trained several.

Exmoor Farmers Livestock Auctions
Cutcombe and Blackmoor Gate

Peter Huntley has been the auctioneer for Exmoor Farmers Livestock Auctions (EFLA) ever since it became a private limited company in September 1997. Its 190 shareholders were mainly local farmers who wanted to save Exmoor's two livestock markets: one at Cutcombe (Wheddon Cross) and the other at Blackmoor Gate. Tom Rook and Robin May both played crucial roles in turning the initial idea into reality.

Sheep and cattle are sold at Cutcombe, but Blackmoor Gate is for sheep only, so the requirements are different at each market site. Various improvements have taken place over the years, including some major building work.

New facilities were completed at Cutcombe in 2010, including a café and offices. The EHFN has an office there, too, and the Market's buildings are used for a range of local events, including Cutcombe First School's annual nativity play.

"It was always our ambition to modernise and update the facilities here at Blackmoor Gate Market," Robin May, chairman of EFLA, said at the opening of the new building on 21st August this year. "But, as with everything on Exmoor, it has taken time!" Robin went on to say it appears there has been a livestock market at Blackmoor Gate for at least 134 years, and that Exmoor breeds both resilient sheep and resilient shepherds.

The sale that followed was a great occasion, with about 1,000 sheep sold. The first lot, a grazing ewe owned by Miss Hazel Scott, was purchased by Patrick Kift for a record price of £250.

"The new building is more user-friendly, especially on a bad day," Peter comments. "The ring is light and airy, and it's much easier to see the bidders."

The number and variety of sales held at both markets have increased since 1997.

"Spreading out cattle sales is especially important so farmers can sell after they've had a clear TB test," Peter says. "TB means farmers often have to sell when they can rather than when they want to."

Cattle numbers at Cutcombe have decreased, with changes in subsidies, TB, poor margins and a shift from quantity to quality all playing a part. Currently about 6,000 a year are sold.

The Exmoor Suckled Calf Rearers' Association sales have become particularly well-known, with buyers coming

Above: "It's the best job ever to be an auctioneer when the trade is good." Peter Huntley selling cattle at the Cutcombe ESCRA sale.
Left: Tracy Parker and Peter Huntley in the rostrum, with Tom Burge's sheep in the ring, at the opening sale in the new building at Blackmoor Gate Market. *(Exmoor Farmers)*

from many parts of the UK for calves to finish. "There are fewer calves now, but the quality has improved no end in terms of conformation and weight," Peter comments. "Quality is far easier to sell. I really enjoy the two-day ESCRA sale. It's the best job ever to be an auctioneer when trade is good, as it has been this autumn."

Blackmoor Gate Market sells about 40,000 sheep a year (although there have been about 5 per cent fewer this year), whereas at Cutcombe it's around 35,000.

Buyers come from all over England and Wales, especially the south of England, with many returning year after year. The traditional role of Exmoor's farmers – breeding sheep and cattle for farmers from more fertile areas to finish – is still important.

EFLA has a dedicated team running it. Three of the founding directors (Robin May, Eddie Schofield in 1997 and George Vellacott in 1998) have been joined by Jeremy Andrews, Paul Nicholas and Tom Burge.

Tracy Parker, who keeps the paperwork and accounts in order, has been there from the start, while the newest member of the team – Hannah Walters from Holwell Farm, Parracombe – became the Environmental Land Management Advisor this September.

Robin is determined the company has a good future, despite current uncertainties about agricultural support and trade deals. "I'm a passionate believer that livestock born on Exmoor should be sold on Exmoor," he says.

Above: Brisk bidding.

Below: The sale ring at Cutcombe.

Jeremy and Angela Andrews,
Lower Woolcotts Farm, Brompton Regis

Jeremy and Angela farm cattle on 600 acres at Lower Woolcotts Farm and 335 acres at Picked Stones, near Simonsbath. Jeremy is Rosanne Coates' brother.

Their children have been brought up with a strong involvement in farming. Harriet (22) is at Harper Adams University studying to become an auctioneer, while Tom (15) is about to sit his GCSEs and is planning to study Engineering at Hartpury.

At one time Jeremy and Angela had over 400 suckler cows and rented additional ground at Emmetts Grange, near Simonsbath, but now they keep about 200 and buy in some additional store cattle.

"This ensures a good supply of heifers to put to the bull each year, keeping the herd young," Jeremy says. "The cattle do better when they're not so thick on the ground, and our farming system is more resilient. The cows have to earn their keep, literally. They're mainly Continental crosses – cows capable of producing a decent calf, with enough milk to feed it – and they're put to a Charolais bull to get even batches of calves that are easy to match in the sale ring."

Above: Angela Andrews at Cutcombe Market.

Calving takes place indoors from the end of March until the beginning of May. "Ideally, the calves go out after a day, but sometimes it's difficult – you can never have enough sheds when the weather's bad," Angela comments. "Cameras in the sheds have revolutionised calving. If cows know you're watching them give birth they get twitchy, but

Below: Angela and Jeremy unloading their cattle for the ESCRA sale at Cutcombe

Above: An evenly matched bunch of suckled calves coming off the weighbridge.

now we can keep an eye on everything remotely from the farmhouse and intervene if there's a problem."

The calves are sold at the ESCRA sales at Cutcombe in the autumn. Jeremy is a Director of EFLA, and he and Angela are shareholders. They say the ESCRA sales provide a very important niche market, supplying cattle finishers from the lowlands with good-quality stock from Exmoor. The same buyers come back year after year, and strong links have been forged.

Healthy livestock with a good shape and weight sell, and that's what they aim to produce. They creep feed the calves from August onwards, which improves growth rates and helps to reduce stress after weaning. "There's an art to suckled calf production," says Jeremy. "Profitability depends on using inputs wisely to get the best possible price at market, but it's a hard balance to get right."

"This year the trade was good," Angela adds. "Jeremy and I delivered some of the calves to Gloucestershire. I like to see our calves safely in their new homes, and I really enjoy driving the lorry and seeing other parts of the country. Harriet has just passed her HGV test as well, so with three of us able to drive we can double-man on longer journeys."

A recent change to their farming system has been the inclusion of herbal leys in rotation with barley. "It's nice to have a bit of variety, and to be able to feed our own barley," Angela says. "There's definitely a balance to be had between the environment and food production, but farmers have got to keep on producing food. Recent policy changes have put the environment at the forefront. However, it's so important that food security should be given equal consideration. As farmers, we're capable of fulfilling both needs."

They have also converted some of their old barns into holiday lets, to conserve the buildings and create another income stream.

"It's funny how things change," Angela remarks. "Farming in a tourist hotspot has its challenges, but now having Wimbleball nearby is definitely an asset!"

Below: Jeremy standing next to the auctioneer, watching his cattle being sold.

Julian Branfield
Mobile Sheep Dipping Service

Julian Branfield has combined farming with contract shearing and mobile sheep dipping for most of his working life. During the past 10 years he has lived in Witheridge, but he keeps a flock of about 170 Exmoor Horns and Exmoor Mules at Knightshayes – an NT property near Tiverton – and about 100 Exmoor Horns at Hawkridge. That's roughly a 50-mile round trip, but he says he's used to travelling as he covers most of the South West with his dipping service.

He is also a staunch supporter of Exmoor Horn Wool, often helping on the stall at shows and doing sheep-shearing demonstrations for the public.

"In a way it's a busman's holiday going to wool events, but I enjoy sheep and talking to people, so that's two of my favourite things rolled into one," he remarks cheerfully.

Below: Unloading sheep for dipping.

"It's amazing how people enjoy seeing sheep being shorn."

Over the years, Julian has got to know a tremendous amount of farms and farmers through his contracting. "I've discovered places I never knew existed, met a lot of different people and seen a wide variety of sheep and farming systems – it's very enjoyable in that respect," he says. "I only shear small flocks of sheep now – I do fewer sheep in the same amount of time – but a girl called Mary Lucas-Ridge sometimes helps me, and she'll be very good at shearing, I think."

In 1992, the dipping of sheep to control sheep scab (which is caused by sheep scab mites) ceased to be compulsory and sheep scab was removed from the notifiable list of diseases, mainly due to concerns about the health risks posed to humans by organophosphate dips. As a result,

Above from left: Sheep in the dip. Sheep coming out of the dip. A good shake. **Below:** All done.

many farmers decommissioned their sheep dips.

"I started with my mobile dipping service two years later because sheep scab just exploded," Julian says. In the first year, he dipped about 60,000 sheep, and the most he ever did in one year was 120,000, but now he does about 30,000 a year. For a while he was the only mobile dipper in the South West.

"There's a lot of putting together and dismantling at every stop – it keeps me fit," he remarks. "I'm lucky because the dip doesn't seem to affect me. In the past I've taken part in trials. The residue used to stay in the fleeces for up to a year, which was truly remarkable, but nowadays, with the modern formulation, it's only five weeks or so."

To control fly strike as well as sheep scab, the dipping season used to be all through the summer and into the winter, but now most farmers use pour-on treatments against fly strike so Julian is busiest in the autumn and winter. "Autumn is the time when most people get scab problems in their flocks because of the autumn sales. Scab is easily transmitted at market, and it becomes more noticeable when sheep are under stress," he explains. "A lot of people dip after they've bought the sheep they want and before the rams go in. Sometimes farmers don't realise there's a problem until later on, though, so I can be dipping right up until lambing. If a flock's got scab, dipping will cure it. Unfortunately, injections don't work as well, so it comes down to animal welfare."

Julian has never advertised – all the work he gets is through word of mouth, and a lot of his customers have become good friends over the years.

"It's not a bad way to make a living," he says, and smiles. "It's kept me off the streets."

The Milton Family
Herd 23

The ancestors of brothers Robin and Rex Milton have farmed in the Exmoor area for at least 500 years. Robin and his family are based at Higher Barton, West Anstey, while Rex and his family are nearby at Partridge Arms Farm. They farm cattle, sheep and arable crops. In addition, they have one of the oldest family-owned herds of Exmoor ponies: Herd 23, with the prefix Withypoole, which Rex manages. Nicholas Milton, who was the tenant of Landacre in the early 1800s, bought some of the ponies from the Royal Forest when it was sold, and kept them on Withypool (then spelled Withypoole) Common. There have been Milton ponies on that common ever since, although they weren't registered with the Exmoor Pony Society until the 1940s.

"The Milton family also had a herd of ponies on Anstey Common," Rex says. "But during the First World War my great-grandfather, James Milton, was the horse procurement officer, and most of the ponies were given away. After the War, he put the remaining ponies back on Anstey Common, but they kept breaking into Zeal Farm, where Charlie Westcott lived (he was Julian Westcott's father). So in the end my grandfather gave them to Charlie, who gave them to his brother Sydney at Draydon, and I believe they ended up at Ashwick."

Fred Milton (Nicholas Milton's great-great grandson and

Below: Fording a river during the Herd 23 roundup. *(Sarah Hailstone)*

Above: Rex sorting out mares and their foals.

James Milton's cousin) inherited Weatherslade Farm from his father in 1946. Fred continued the family tradition of keeping ponies on Withypool Common until his death in 1998. As he had no children, Weatherslade and the pony herd passed to Robin and Rex.

Anstey Common was restocked with Withypoole ponies eventually, so the two herds have the same herd number and all the ponies are closely related even though they graze two different commons. There are about 40 mares in all: 25 on Withypool (which amounts to about 3,000 acres and is owned by Landacre) and 15 on Anstey (900 acres under multiple ownerships).

"Everyone used to use their stocking rights much more," says Rex. "For instance, Uncle Fred used to have about three hundred ewes, forty cows and forty Exmoor pony mares on Withypool Common, but now we just graze our twenty-five ponies, and other farmers graze an agreed number of cattle and sheep at certain times of the year."

Because of the lower levels of stocking, the ponies thrive and are much easier to manage because the Miltons are effectively the only graziers with breeding ponies on both commons.

Unlike sheep and cattle, the ponies are allowed to graze the commons all year round. "They thrive on moorland, and the moorland benefits from being grazed by them, so we've got to keep that going," Rex says. "You don't see cattle eating gorse very often, do you? But the Exmoor ponies are always nibbling away at it. The foals, especially, love eating gorse flowers."

This is the first year Rex can remember foals coming in with lots of ticks on them, and he considers that's largely because the burning of moorland (or swaling, as it's known on Exmoor) isn't allowed under HLS agreements. He echoes the opinions of many moorland farmers when he says: "The basics behind the environmental schemes for commons have been good, but they should have got eighty-year-old farmers with a lifetime's experience of moorland management to make the rules rather than people fresh out of university."

Having taken part in the roundups for 48 years, Rex has got a wealth of experience and has developed an instinct for where the Exmoor ponies will go and what they'll do.

"I ride a quad bike rather than a horse nowadays, but my wife Banger and daughter Rosie ride horses on the roundup, together with six or seven others. It takes a good horse to keep up on Withypool Hill, especially. The speed at which Exmoor ponies move is incredible, and they know every inch of the moor."

Over the years, the market for ponies has changed. Fred Milton used to walk the ponies he wanted to sell to Bampton Fair each autumn, and Rex can remember helping him. However, since the 1980s, the Miltons have sold their surplus ponies privately. "Last year we sold sixteen yearling fillies to the Purbecks as ecological grazers, which was a really good sale, but this year we've lost out because a major rewilding project was postponed due to uncertainty about the new schemes and whether there'll be a native-breeds supplement," Rex says. "The market's there for a sensible number of ponies, but it's ridiculous to breed too many. Prices for ponies have been much better in recent years, and the Exmoor Pony Society, Exmoor Pony Centre and Moorland Breeders' Group have all played a part in that. We've got to work together to look after the ponies for their long-term future; Exmoor's moorland and its ponies are part of our heritage."

Below: Robin reading microchips in the ponies' necks with an electronic stick reader.

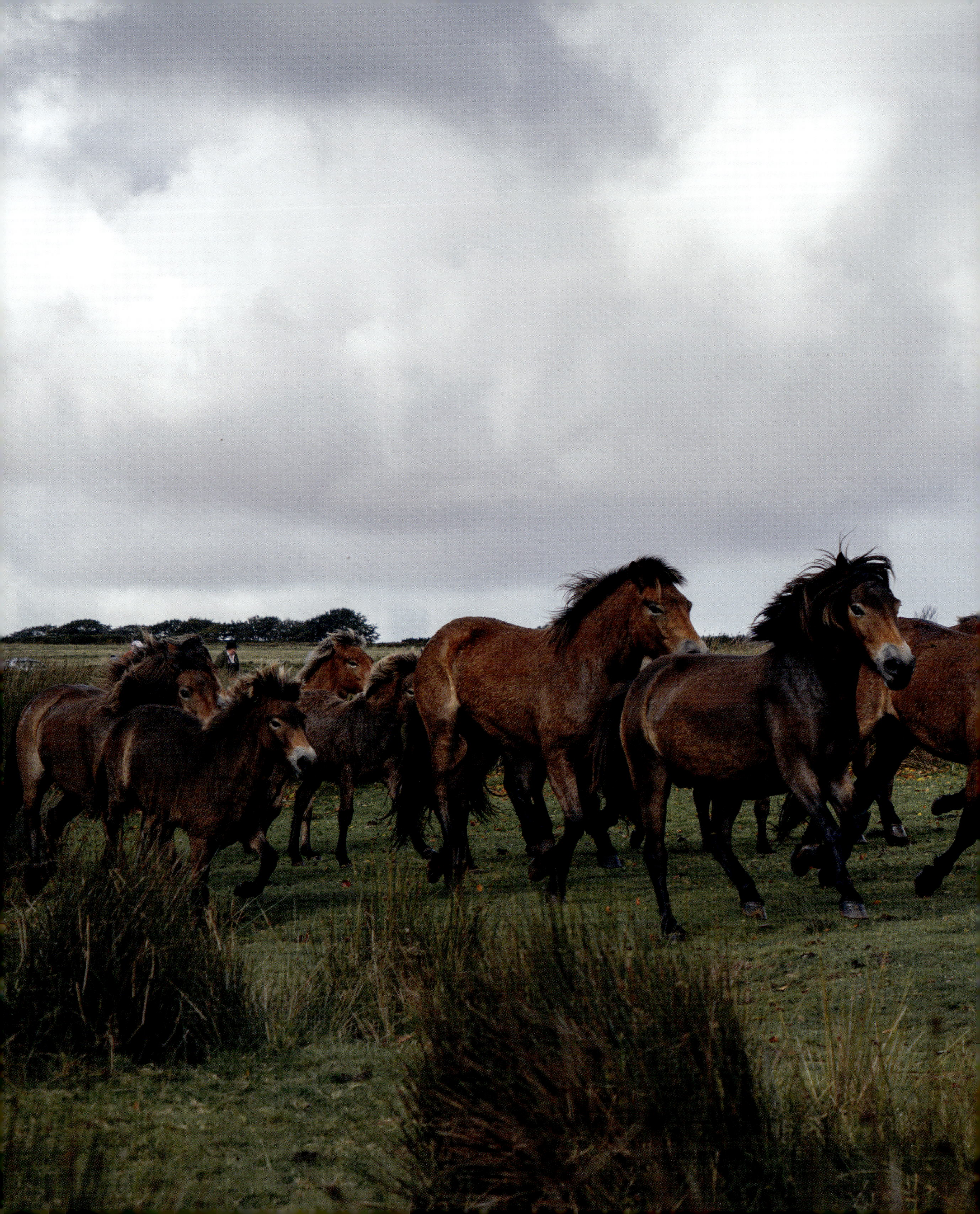

Gathering Exmoor ponies on Bradymoor. *(Sarah Hailstone)*

The Wallace Family
The Anchor Herd

In the late 1920s, an industrialist from Yorkshire called Frank Green bought Old Ashway Farm and its moorland allotments of Ashway Side and Varle Hill, together with some of the Exmoor ponies that grazed there. He also bought Ashwick House, and ran the Ashwick Estate from there. In those days, there were about 20 resident staff in the house.

To provide for the entertainment of both staff and visitors, a wooden building known as the Music Room, with a dance floor and stage, was constructed in the grounds. The building has recently been restored and is used as a tearoom and meeting-place by the Exmoor Pony Centre, which rents land and buildings on the Ashwick Estate.

Other Exmoor ponies came from the Acland family (whose ponies had a distinctive anchor brand) and the Westcott family at Draydon Farm. Soon the anchor brand was given to all the registered Exmoor ponies born on the Ashwick Estate, whether they were grazing the Estate's privately owned moorland allotments or Winsford Hill, and the herd became known as the Anchor Herd.

During the Second World War many ponies were taken for meat, and one night most of Frank Green's ponies were stolen. About a dozen ponies that had evaded capture were brought in-ground to the safety of the farm for a few years.

In 1947, Old Ashway passed to Frank's great-nephew, Simon Green. Rosie (Simon's daughter) married Ronnie

Above: Emma getting Anchor Moonstone ready for the show before David takes him into the Exmoor Pony Stallion class at Exford Show. *(VE)*

Below: David leading Anchor Moet in the Dry Mare class at Exford Show. *(VE)*

Below: L-R: Emma, Anchor Rock Dove, David and Anchor Moet at Exford Show. *(VE)*

Wallace in the 1960s, and they moved to Mounsey Farm. By that time, she was looking after the Anchor herd. Rosie was very keen on showing, and her successes in the show ring helped to make Anchor ponies sought-after. After she died in 2005, her son David and his wife Emma became custodians of the Anchor Herd.

Today there are around 100 ponies in all, including stallions and youngsters. There are three senior stallions (Anchor Moonstone, Anchor Monster Munch and Anchor Labyrinth Lord), two two-year-old colts and four young colts.

"We've got bloodlines from all eight founder mares, and we're working to preserve them through careful breeding," Emma explains. "The mares generally stay out for most of the year – about thirty-five on Winsford Hill and the rest on the moorland allotments – and then they're rounded up in October for weaning and the Exmoor Pony Society foal inspections. An exception is the Mercury family, which is kept in-ground because there are only three ponies left from that bloodline."

The road over Winsford Hill makes grazing it hazardous. "About one pony is killed there every year," Emma says.

"Anchor Fairytale went to the Horse of the Year Show in 2016, where she was the highest-placed Exmoor and homebred pony. After that she had a lovely filly foal on Winsford Hill, but unfortunately her foal was killed on the road."

Margaret Rawle is a councillor in Dulverton, and she is helping to make drivers more aware of driving carefully where livestock are grazing.

Like Rex Milton, Emma thinks that controlled burning is a good way of encouraging heather and decreasing the tick burden on moorland. The Wallaces undertake controlled burning on their own moorland, and the heather is coming back. "Touch wood, the ponies don't seem to be too badly affected by ticks, but we did lose a mare to red water fever a few years ago," Emma comments. "We don't put cattle out on moorland because of the tick problems."

After the foal inspections and registrations, the weaned foals are usually kept in stables for a week or so for basic handling, which makes them easier to manage.

"Our groom, Sophie, is a really good girl, and Andrew Hagley (who has been on the farm for forty-two years) is brilliant with the ponies," Emma says. "Staff from the

Below: Anchor ponies crossing the road.

Above: Coming home.

Exmoor Pony Centre sometimes help people who have bought our ponies with their initial handling, and recently they have also helped with a couple of our show fillies. The Centre is a great asset, and it's become quite a tradition that ponies and riders from there help to gather Winsford Hill."

One of the purposes of the Centre is to take Exmoor ponies for handling that haven't been sold from the moorland herds, to help secure good homes for them. Some become ecological grazers while others become family pets, and a few are kept. For the past few years, though, the surplus Anchor foals have found new homes soon after weaning, with buyers travelling from all over the country and overseas. Anchor ponies have been sold to the Czech Republic, Canada, the United States of America, France, Germany and Sweden.

"We usually keep all the females until they're three years old so we know exactly what we've got and can choose which ones to keep in the herd when they've matured," Emma explains. "It's lovely when people buy our ponies and do well with them in the show ring. This year Anchor Jaffa Cake and Anchor Lullaby went to the Horse of the Year Show."

Emma's background is in show horses, and she's a judge for the British Show Horse Association. She judges Exmoor ponies and mixed classes, too, and David is an Exmoor Pony Society inspector.

"There's not much difference between judging Exmoor ponies and other breeds," Emma remarks. "I look for correct conformation and movement, with the breed type on top."

The Wallace's children, Miranda and Lawrence, have been brought up with farming, Exmoor ponies and showing, and Lawrence intends to be actively involved in running the Estate and the Anchor Herd.

"At the end of the day, Exmoor ponies are a vital part of our heritage," Emma says. "They shape Exmoor."

Woolhanger Sheepdog Trials

The annual two-day sheepdog trials at Woolhanger every autumn has gained a reputation for being one of the toughest in the country. It was started about 20 years ago by David Kennard, who organises it with great support from the Woolhanger Estate.

David farms near Mortehoe, on the North Devon coast. He is a well-respected farmer and has shared his love of sheepdogs with a wide audience through his books and the ever-popular TV series *Mist: Sheepdog Tales*, which was filmed in North Devon and on Exmoor.

"Woolhanger is such a different trial. Some people travel a long way for it. For instance, this year Katy Cropper came from Cumbria and Rob Ellis came from Wales. The Welsh, especially, seem to enjoy coming down," David says. "The layout of the course, with a steep valley and stream in the middle of the drive, makes it particularly testing, and the Cheviot gimmers at Woolhanger are good at finding weaknesses. There's an awful lot that goes on between dog and sheep; a dog has got to have presence to control them."

Above: David Kennard and Mirk. *(VE)*

This year, Rob Ellis and Fly won on Saturday, and David won on Sunday with his dog Mirk. Katy Cropper and Gin came third on Sunday, and Tim Nicholas came fifth

Above: David Kennard says, "The cross-drives are particularly tricky at Woolhanger because they are so far away and it's very difficult to judge where the sheep are in relation to the gates - until you've missed!" *(VE)*

with Todd. (Tim gave the sheepdog demonstration at the Closewool Centenary in August.)

"It's lovely to see Tim doing so well," David says. "I helped him a little when he first started training sheepdogs. He has a lovely quiet way with his dogs; he's very talented."

Another competitor who David praises is Debbie Survila from the Blackdown Hills, who competed with five-year-old Windcutter Mia (bred by David) and eight-year-old Sanduck Jock. "It's really impressive to do so well when she hasn't been trialling for long and she lives on a smallholding," he says. "Neither dog was placed but they completed their runs, and only about a third of the competitors manage that."

"I've bred rare-breed sheep for thirty years, and I've had dogs and horses for a long time, but I always thought sheepdogs were so special that you really had to be born into that world in order to be able to work with them,"

Debbie admits, as she sits on the hillside with her dogs, watching her fellow competitors. "So, this all started with a love of sheep; it helps to love sheep if you plan to work and train a sheepdog, I think. It took a lot of bravery to buy my first-ever sheepdog, Jock. He wasn't straightforward, and to begin with I couldn't find anyone to help me. We had to learn together."

Mia pushes under Debbie's arm, and she strokes her. "I liked Mia as soon as I saw her. I've always admired David's dogs and how incredibly versatile they are, so I feel very privileged to have a Windcutter pup."

Although dogs have always been part of Debbie's life, she says working a dog with sheep is totally different. "The feeling of running my dogs around this course is second to none. It's hard to put into words. It's a timeless connection through the heart, standing on the earth and experiencing such a deep working relationship with another animal."

Below: Debbie Survila and her two sheepdogs watching a fellow competitor completing a cross-drive on the other side of the cleave. *(VE)*

Rabbit Slattery
Wootton Courtenay

"As a kid I used to watch *One Man And His Dog* with my family, as we were all very animal-orientated, and I suppose that's when my interest in sheepdogs started," says Rabbit Slattery. She is the youngest child of the legendary equine and human physiotherapist, the late Mary Bromiley.

After training as a sports therapist, Rabbit worked in racing for 31 years, but she now leads a quieter life at Wootton Courtenay, where she keeps Soay sheep on about five acres. She also has several sheepdogs.

"My first dog was Bob," Rabbit says. "I bought him just before I moved down to Exmoor in August 2000. Mum had already moved to Combeleigh Farm, near Wheddon Cross, and had a flock of sheep. I bought a book about training sheepdogs, and it advised building a round pen so the sheep would be inside and the dog could run around them outside. It took me three hours to get the sheep inside the pen, then Bob jumped in and the sheep jumped out! He ended up being a good working farm dog, though."

Her next dog was Saffi. Rabbit bought her as a puppy from a good trialling family. "I was determined to start her properly, so I enrolled for a six-week course with a sheepdog trainer called Roderick Hayes. That set us on the right track. Saffi was difficult but a natural, and we gradually went up the ranks at sheepdog trials. She took me to the Nationals three times."

Rabbit has also competed at Woolhanger three times with her current best dog, Lash. She bought him through renowned sheepdog trainer Ricky Hutchinson, who owned the sire. She met Ricky at the Nationals and invited him down to Combeleigh to do some sheepdog training clinics for her and other owners.

"Lash is possibly the best dog I've ever had, but he's getting quite old now. Various things – my illness, Mum's death and the Covid pandemic – all meant I didn't do as much as I would have liked with him," Rabbit says. "His half-sister Anya became a good farm dog, and I've got Arie, one of her puppies."

Rabbit really enjoys the puppy stage. All her puppies start

Above: Rabbit with Lash

off in the house, but learn to be kennelled during the day, and they move out to their own kennels at about six months old, or after their first winter. She teaches them to walk on a lead, sit, lie down and come when they're called before they start working.

"It's really satisfying taking a puppy right through to trialling. Winning a prize is great, but for me it's about working well with my dog," she says. "Each dog is an individual, and you have to train them all differently according to their strengths and weaknesses. I love training, but I don't feel I've got enough experience to teach sheepdog handling. When you see Ricky Hutchinson and his wife teaching, you realise what a skill it is – they're exceptionally good at reading the handler and the dog together."

Rabbit has just had a litter of puppies by Lash. "At the moment I don't have a dog to trial, but I would like to again in the future... I'll probably keep a puppy."

Left: Two of Lash's puppies.
Below: Lash rounding up the Soay sheep.

Harvest Celebrations

Winsford

During the autumn, harvest is celebrated across Exmoor's parishes with services of thanksgiving and harvest suppers. 'Harvest' for Exmoor's livestock farmers generally means making hay and silage, rather than harvesting grain, but in the broader sense it also means weaning and marketing livestock.

On the 6th October, 60 people sat down in Winsford Village Hall to a delicious supper prepared by a large team of dedicated helpers. This autumn the hedgerows provided a bumper harvest of blackberries, and fruit crumbles sat pride of place amongst the desserts.

Richmond Harding, a popular local farmer and livestock judge, took charge of the auction which, together with the raffle, helped raise a healthy sum for Winsford's St Mary Magdalene Church and the local food bank.

Above: L-R: Tish Brown, Richmond Harding, David Weir and Jane Pearn. *(David Butt)*

Blackmoor Gate

On the western side of Exmoor, on Sunday 15th October, parishioners of St Thomas' Church, Kentisbury, and several local farmers gathered together in the new sales building at Blackmoor Gate Market for the first-ever harvest festival there. It was the idea of Market Chaplain Andy Jerrard, and everyone agreed it was a huge success.

The Reverend Prebendary Rosie Austin took the service, and Odette Dunn (who farms at Valley View, Kentisbury, with her husband Andrew) played the keyboard. In his address Andy thanked farmers for the work they do, which was very much appreciated.

Patrick Kift was the auctioneer for the evening, and dealt admirably with fierce competition for several lots – especially the pumpkins and chutney.

Afterwards, everyone enjoyed a delicious supper.

Far left: Andy Jerrard, Rosie Austin and Odette Dunn. *(VE)*

Left: Patrick Kift selling the last jars of chutney. *(VE)*

In 2014, at 20 years old, he won the Exmoor Society Pinnacle Award. "I bought a post banger with the money, but the best thing about winning the Award was the publicity," he says. "The phone never stopped ringing; it definitely got my name out there."

By 2016, he was contracting full time with a digger and two John Deere tractors – doing a lot of fencing and tractor work.

Fran met Jack in 2015, when she was 18 and he was 21. "I wasn't particularly into farming but Jack got me into it," she says. She has learnt to operate all sorts of machinery and has become, Jack admits, the best hedgelayer in the team.

By 2017, she was working full time in the business, and now she's Jack's business partner and fiancé. They employ a bookkeeper and accountant, but she does a lot of the day-to-day administration.

Above: Will Hunter thinning out a hedge with tree shears to make it easier to lay.

Below: Partially cutting a stem (known locally as a steeper) on the diagonal to leave a hinge of living wood so it can be bent over and laid (or steeped down).

Above: Steepers are bent over to form a living stockproof barrier. Regrowth will occur from the cut base and from the steepers.

Above: A peg-shaped piece of wood, called a crook, is cut from a tree and driven into the bank to secure the steepers.

Above: Jack steeps down as much as possible to create a thick, stockproof hedge that will grow well.

"Even making something to eat is a struggle when we've been out working all day," Fran says. "We often fall asleep while we're eating our supper. And keeping up with all the rules and regulations is very time-consuming."

This autumn and winter, their main jobs will be mires restoration, fencing and hedging.

"We've got three lads doing mires restoration work until March, when we have to stop due to ground-nesting birds," Jack says.

They are currently re-wetting 60 acres of ground on Shoulsbarrow Common, through the South West Peatland Partnership, for Peter and Dianne Wyatt from Buscombe Farm, near Challacombe. The Wyatts say they were in two minds about the project, but they had to agree to it to qualify for a Higher Tier CS agreement.

The idea is to stop the erosion of peat, bring back sphagnum moss and get the peat growing again. Amongst other things, Jack and his team will install hundreds of wooden dams, plant trees like alder and willow and erect deer fencing.

"The business does about seven thousand metres of hedging a year, plus thousands of metres of fencing," Jack says. "Many more people are cutting and laying hedges these days. There's a lot more in terms of grants to help with the cost now, and tree shears have made the job much easier."

He and Fran have achieved a huge amount in the past few years. Their plan now is to consolidate what they've got, continue to provide their customers with the best possible service and put a roof on their house before the next band of winter weather rolls in.

Above: A spirit level is an essential tool.

Above: Building wooden dams for a mires restoration project on Shoulsbarrow Common.

Above and right: A series of wooden dams stop the erosion of peat by curbing the flow of water downhill – allowing silt to build up and new peat to form over time.

Marthe Kiley-Worthington
Cranscombe Cleave, Brendon

"We have been at Cranscombe for only three years but we are gradually developing this small forty-two acre farm as a food-producing nature reserve and educational centre," Marthe Kiley-Worthington says. She and her partner Chris Rendle have been practising 'ecological agriculture' for 45 years, which they define as diversified, self-sustaining farming that is economically viable and has a good carbon and energy balance, where all species have a life of quality and the whole farm is also a nature reserve.

"Food must be produced more efficiently than it has been to date, in terms of both energy budget and carbon footprint," Marthe insists. "If an ecosystem is energy balanced it is stable; if out of balance, it changes."

Marthe also emphasises that species diversity is important:

"The beautiful living world must be able to express itself. We need different eco-zones on the farm – ponds, muck piles, loamy beds, grassland, woodland, gorse, nettles, brambles, shrubs, fruit trees, flowering plants, wet areas, warmer areas and so on. Each species, whether indigenous or introduced, has interwoven effects on the whole ecosystem, and produces something for the ecosystem and for us humans."

Some previously grazed land at Cranscombe has been set aside for woodland and scrub development to encourage more wild species into the pastures, which are managed organically. "We have wet land, leaky dams, old hedges and new hedges, orchards, vegetables and two greenhouses. And we will have a small area of cereals, plus a dairy so we can produce milk, cream, butter, yoghurt and cheeses from one or two cows," Marthe says. "We've also got

Below: Marthe and Chris using horsepower to cultivate the soil. *(VE)*

Above: Open day at Cranscombe Cleave. *(VE)*

poultry: hens, ducks and geese. It's hoped that, as well as grazing the grass, the geese will protect the hens from marauding foxes."

Earlier in the year, Marthe tamed a few previously unhandled Exmoor ponies from Champson Farm and took them to a nature reserve in France for a rewilding project. She is also acquiring red and fallow deer for the nature reserve.

Another project she is working on is the completion of a reference library for the local community, with books on agriculture, equines, ecology, ethology and animal minds. She is a trustee of the charity, We Are All Mammals.

Marthe and Chris are firm believers that the physical and mental needs of all the animals in their care should be met. "This does not mean that they can do exactly what they like when they like," Marthe remarks. "But they must be able to exercise choice and have the freedom to perform all the behaviours within their repertoire, as long as that doesn't cause suffering to others."

They breed cattle and horses, and use them to provide energy for appropriate work on the farm, like harrowing and cultivating. Animals enrich their lives as friends, as well as benefitting the ecosystem, sequestering carbon, working on the farm and providing food. Visitors to a farm open day in August were able to look around their farm and see their horses in action. This was when the photographs were taken.

Marthe thinks that the Exmoor National Park could be a model for the future of farming and food production. She and Chris are working hard to show it's possible to integrate wildlife and diversity while producing a sustainable surplus of food for others to buy.

Above: Young geese relaxing in the garden. *(VE)*

Holly Purdey and Mark Brewer
Horner Farm, Porlock

When Holly and her husband Mark took on the tenancy of the NT's Horner Farm in January 2018, their son Fergus was a year old and they had 150 ewes.

They've now got three children: Fergus (6), Moss (4) and Lachlan (6 months), and they're farming sheep, Boer goats (for meat), cattle, pigs and chickens.

"We're aiming to combine positive ecological practices with food production and humane livestock husbandry. Keeping a variety of grazers and browsers on the farm helps us to achieve that," Holly says. "Our goats, for instance, are brilliant at browsing weeds like docks, thistles and nettles, while goldfinches love the thistles we leave for the goats. And the pigs are good ecological disrupters; we spread wild-flower seeds behind them. If there's a natural process that can deal with a problem, we'll try it – like including trefoils in our herbal leys to act as natural wormers and reduce ruminant methane production."

Connecting people with the food they eat is very important to Holly and Mark. Their meat is sold through their farm shop and through Holly's brother's shop in Wellington, they host a monthly producers' market at Horner, and they have special feast nights and pizza nights at the farm. Educational and private farm visits can be booked, and they also put on events with a family focus, like lambing experiences and, recently, a family astro-party with astronomer Jo Richardson for the Dark Skies Festival.

A three-acre field on the north side of the farm is used by friends Adam Reed and Leighanne Beart for a community-supported agricultural group called Good Vibe Veg, which yields about 35 boxes of vegetables a week that are delivered to local drop-off points. Their produce is also used for feast nights and other occasions.

Below: Boer goats originated in South Africa, where they were selectively bred to thrive on thorny vegetation and produce good quality meat. They are good at keeping weeds like thistles and stinging nettles under control. *(Holly Purdey)*

Above: Sheep grazing above Horner Farm.

Above: Holly and Lachlan with some Shorthorn cattle.

Above: Good Vibe Veg is a cooperative venture at Horner Farm.

The integration of trees is a particularly important aspect of the work being carried out at Horner. "Trees provide shade and shelter, are good for soil health and provide livestock with browse," Holly explains. "Mark is a tree surgeon, and he has been the instigator of many of the changes we've made."

They have planted areas of silvopasture (rows of trees in species-rich pasture), low-density wood pasture (trees dotted around on about 20 acres of sloping pasture) and also individual fruit trees. The planting of trees in field corners and carefully targeted areas to slow the flow of water is also taking place.

Some of their hedges have been fenced to allow them to widen and create wildlife corridors, and in other places new hedgerows are being planted.

"I want to be able to manage the livestock on a rotational grazing system using hedges rather than electric fencing. Hedges give all-important shelter, amongst other things, and I really don't like electric fencing," Holly says.

Last year, a bid by the NT for an ELM Landscape Recovery Scheme, which includes Horner, was successful, and they're now in the two-year development phase.

"It's definitely a positive for our farm; we'll be able to do a lot of the things we want to do for the environment at the same time as producing food," Holly says. "However, I'm particularly uneasy about private deals for things like carbon, as they create credits out of thin air that others can buy to offset the harm their businesses are doing. That isn't going to save the planet – in fact, it could penalise farmers who have been farming responsibly all along because they can't add as much. It seems to be a minefield, and it's a distraction from what we could be doing as farmers."

Exmoor Vintage Beef
at Horner Farm

On a stormy evening in mid-November, local chefs gathered at Horner Farm to sample some inspired dishes created by The Wild Pear Catering Company using 'vintage beef'.

Exmoor Vintage Beef is the brainchild of Vickie Ward from Easterdown Farm, Tara Wright from Castle Hill Farm and Holly Purdey from Horner Farm, based on their desire to be responsible for all the animals on their farms and minimise waste. They are hoping to recruit a network of Exmoor farmers who are willing to offer their older cattle for direct sale to local catering businesses.

Georgie Everett, who runs The Wild Pear, was brought up in Exford and is passionate about cooking with locally produced ingredients. "I really enjoyed cooking the vintage beef – it has a tremendous depth of flavour," she said after the event. "And I love the idea of keeping things local and using as much of every animal as possible."

She does the cooking for Horner Farm's popular feasting nights, and is hoping to be able to incorporate vintage beef into her menus in future.

"My drive is connecting people with food," Holly said to guests at the event. "The stories of the farms supplying the food we eat should be told."

Above: L-R: Vickie, Tara and Holly.

Above: Tara talking passionately about Exmoor Vintage Beef.

Above: Georgie Everett, who runs The Wild Pear, cooked some delicious food for the guests.

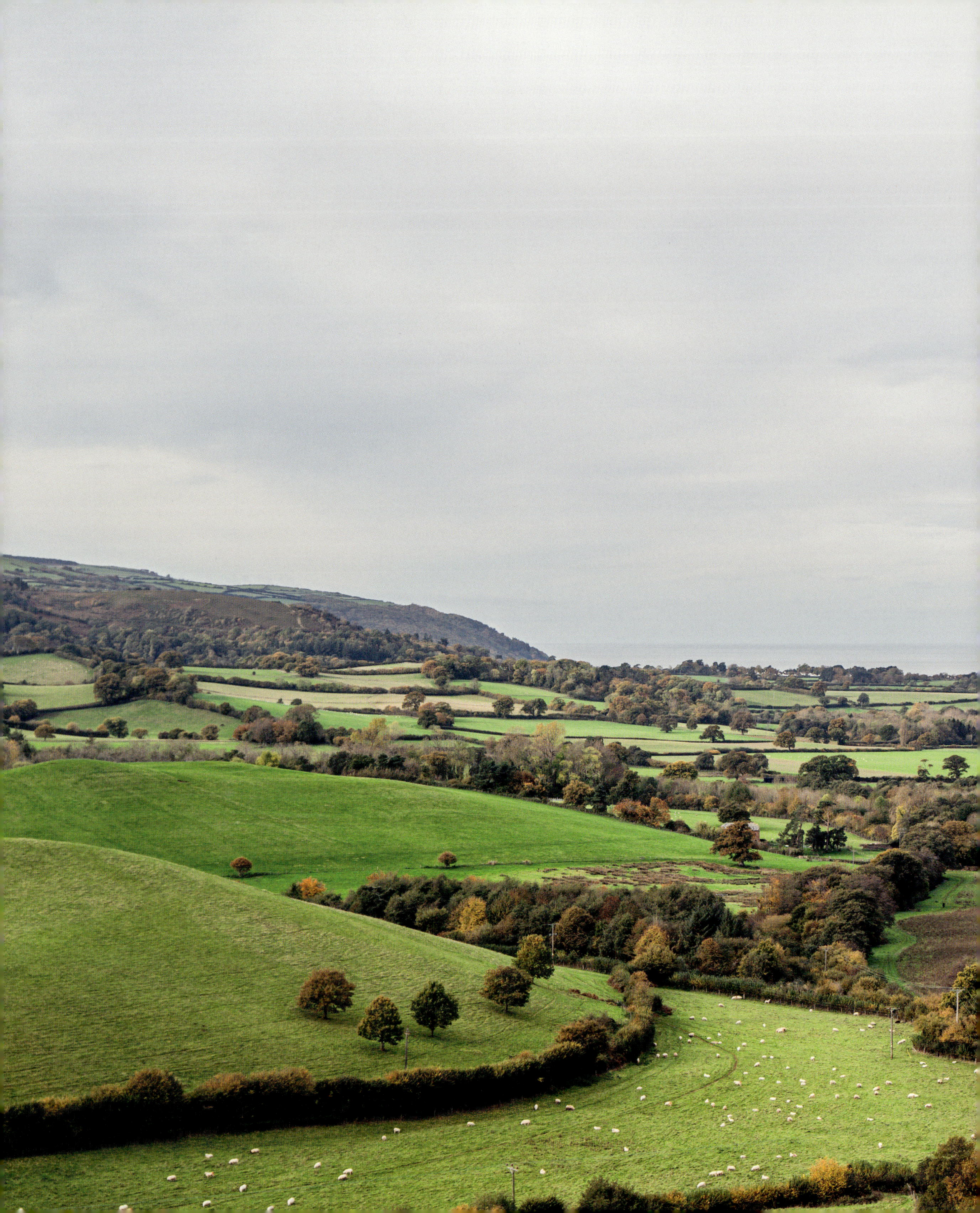

Porlock Vale from the road above Wootton Knowle.

The Hollam Estate
Dulverton

"What we're really proud of is the work we're doing for conservation," Annie Prebensen says. "We have three interwoven strands to our life here – nature, the farm and the shoot –and our aim is for them to sustain each other."

Annie moved to Hollam 20 years ago with her husband Preben Prebensen, drawn by its beautiful landscape and wonderful trees.

The Mildmays, who owned the Estate for about 300 years, did a lot of planting of both native and ornamental trees, including palms. "Another big feature of Hollam is the Scots pines," Annie says. "We've planted lots more, and they're doing brilliantly. The ones we planted eighteen years ago are twenty-five feet tall now. We've also planted species like oak, lime, sweet chestnut, wild cherry and beech to replace those specimens lost through age. We found a map in the attic showing how the Estate was in the nineteenth century, and it's amazing how many more trees there are now."

Over half the Estate is covered in woodland, and new plantations have been added in the past 20 years, including one with wild-flower mix underneath the trees called Frankie's Wood, which has been created in memory of Annie's mother. The ancient woodlands are managed in various ways, such as thinning, coppicing and clearing invasive species like rhododendron. Other small areas are left unmanaged.

A 30 metre-wide wildlife corridor has been created between the Little Exe and Barle Valleys, planted with wildlife mixes as well as trees and shrubs. Elsewhere, hedgebanks are cut and laid in the traditional manner every 10 to 12 years. Over 1,250 metres of new mixed native hedges have also been planted, and in-field trees have been added.

Ponds and watercourses have been improved to increase their wildlife value and minimise runoff, and at least 50

Above: Annie Prebensen. *(Hollam)*
Below: Staff and helpers on a shoot day. *(Hollam)*

Above: Rare plants and insects have been found in the valley leading to the house.

Above: The ancient woodlands are steep and have many springs and streams.

Above: Hollam House.

acres of grassland are now managed as wild-flower meadows.

The farm livestock at Hollam are seen as valuable land-management tools as well as providers of meat and an income. There are around 65 suckler cows, which are put to Charolais bulls, and 250 Exmoor Horn Mule ewes, which are put to Texel rams.

"We try to grow all our own food here; we haven't bought meat for twenty years," Annie says with pride. "J C Le Grande, who is the chef for the shoot, is a genius and is committed to using what we grow."

Hollam is an important part of the Dulverton community. The team includes four families that live and work on the Estate, together with others who live locally, and on shoot days about 25 additional people are employed.

Pete Stanbury is the woodland, farm and Estate manager, and he and his father Rick also help to host the shoot; Paul Beer is the livestock manager; and Liam Curtis, the head keeper, works with Mark Smith to manage the shoot as sustainably as possible. The stocking density of pheasants is kept relatively low, shooting is limited to 38 days, only lead-free cartridges are permitted and cover crops are entirely perennials sown with wildlife mixes.

"Fortunately, the things that are good for pheasants are also good for wildlife," Annie comments. "Our game plots are wildlife havens."

Hollam is taking part in a NE survey on the effects of releasing pheasants into woodland, as well as other wildlife-monitoring programmes with ecologists like Dave Boyce.

"I'm increasingly fascinated by the wild environment," Annie says. "It's so important to have diverse habitats – not just trees. Skylarks, for instance, need space. Diverse, linked habitats support biodiversity. Rare dormice, lichens, bats, butterflies, beetles and plants have been found here. We're all committed to improving the Estate for wildlife. It's amazing how nature thrives if you give it a chance."

The Woollacott Family
Oareford Farm, Oare

Brian and Geraldine Woollacott met on the school bus, travelling to and from Oareford and Brendon Manor respectively. They married in 1974, settled down at Oareford and had three children: Julia, Sarah and Jonathan (Jono).

"I wouldn't say farming's fun, exactly, but it's a way of life," Brian says. "I used to love riding and hunting, and I did a lot of horse breaking with Father. Things didn't change as fast as they do now, and there wasn't the red tape and paperwork."

Below: Jono with his dogs. L-R: Rex, Jet and Bruce.

Brian's way of life at Oareford was passed down to him by his father Jack and grandfather Robert, and now he's gradually passing it on to Jono, who farms it with Oliver Lowday, the full-time farm worker.

Like many farmers, Jono has reduced the number of livestock and the amount of bought-in inputs like feed and fertiliser. They now keep around 90 cows and 1,500 ewes.

"We're calving inside from January to April," Jono says. "Our cows are Hereford cross Friesians put to a Charolais bull – not the hardiest, but they produce good suckled calves to sell at Cutcombe."

About two-thirds of the farm is moorland. Cattle graze there during the summer, and a flock of Scotch Blackface ewes lives there all year round.

Jono recently changed the farm's main flock from Exmoor Horn to Cheviot cross ewes. "They're hardier, and it may work okay to lamb them outside," he explains. "Lambing indoors is pretty labour intensive; we employ a night lamber plus some students. About fifty to eighty ewes lambing each day is good, but I remember having nine hundred in seven days once!"

The rams are put with the ewes towards the second week of November so lambing will begin just after the beginning of April. Raddle is used to show when the ewes have been mated. This means they can be split into different groups according to when they're likely to lamb, which saves on shed space, feed and manpower.

"We don't put any raddle on the rams for the first five days, then there's a different colour every five days for fifteen days, and one more colour until the rams are taken out," Jono explains. "We used to paint raddle powder mixed with oil on the rams' chests. It makes a good mark on any ewe that's mated, but it was one hell of a job to catch up to sixty rams every couple of days to paint their chests! So now we use raddle harnesses with solid crayons of raddle. They don't mark quite so well, but they save a lot of time and reduce stress."

Jono also works as a shooting instructor, especially during the autumn and winter.

"When I was about ten years old, I was given a four-ten

shotgun, and that's how I started shooting," he remembers. "I did shooting at school, too, and eventually helped out with teaching others."

After leaving school, he bought shooting days at local auctions whenever he could. A few local people asked him to help them with their shooting, and he found he really enjoyed teaching. He did an instructing course, bought some clay traps, and a hobby turned into an enterprise.

"I'm away a lot – I've even been to France and Spain teaching – and I've met some amazing people, but I do feel guilty about being away from the farm," he says. "In the long term, I don't know what the future will bring. I'm very confused about the direction of farming right now, but hopefully the way forward will become clearer in the next year or two. We've got to be willing to adapt and make the most of opportunities."

Above: 'I bet you say that to all the girls.'

Below: Rounding up the sheep so that the raddle harnesses on the rams can be checked, with Hookway Hill in the background.

"The pig enterprise is our most recent one, and arguably the most successful," Anne says. "We buy weaners every month from Dennis Ballard and Beth Hallam a few miles away at Highley Farm. Our pigs are inside during the winter, with silage to eat and wool to lie on, but for most of the year they're outside. We graze them on the rough areas of the farm, moving them on before they damage the soil. They adore brambles, and also love pink campion and sorrel."

There's a huge difference between intensively reared pigs and those that are kept more naturally – both in the story behind the meat and the quality of the meat itself – so the pork from A May's Farm is in hot demand.

The suckler cows at Slattenslade used to be mainly Black Hereford cows put to a Limousin bull, but since 2015 Alex has been using Stabiliser bulls, working towards having a herd of Stabiliser cattle. The Stabiliser is a composite breed, developed by scientists in America using several different breeds carefully selected for specific traits with the aim of producing the world's most efficient suckler cow that's docile, polled, fertile and easy to calve, with good growth rates, feet and milk production, plus high-quality meat.

"We really enjoy having Stabilisers," Alex says. "The Slattenslade herd calves outside in the spring, then the yearling cattle go to Dad at Limetree for finishing."

During 2020, Alex and Anne reared 24 Black Hereford heifers as a lockdown project, and 10 of the best became the foundation cows for their autumn-calving herd at Silkenworthy, along with four pure-bred Stabilisers. Their cows are all put to a Stabiliser bull, so they are working towards having a herd of pure Stabilisers there as well. The male calves from Silkenworthy are finished on-farm and sold as beef under the A May's Farm brand.

Finding a breed of sheep that suits them hasn't been so easy. Initially they had Exmoor mules at Slattenslade, but replaced them with Highlanders – a modern composite

Above: Young farmer George. *(Anne May)*

Above: Outdoor pigs at Slattenslade. *(Anne May)*

Left: Young farmer Lily.

breed developed in New Zealand. As part of his degree course, Alex spent a year working on farms in New Zealand, and that influences how he farms now. For instance, rotational grazing is popular there, and he and Anne have found it helps grass growth, biodiversity and soil plus animal health at home. However, Highlanders didn't perform well at Slattenslade and couldn't cope with the rain, so they're being replaced with a flock of hardy Swaledale ewes that can lamb outside in March.

The Mays have lamb, pork and beef available to buy all year round from the freezer, and they have fresh meat for the Lyn Valley Market on the first Saturday of every month. Anne also sells meat every Tuesday outside Lynton Town Hall, to coincide with the fish van. "We both help each other. We'd love to get a market gardener involved, too," she says. "It's so satisfying to stand by our meat, and very gratifying when people say how much they've enjoyed it. Local food for local people is key. Our local abattoir is crucial, too. It takes less than thirty minutes to drive our animals to Combe Martin Meats, and the staff are so professional."

Farming at Slattenslade has given Alex and Anne a fantastic foothold on the farming ladder, but long-term they'd love to build their own house at Silkenworthy and farm in their own right. "For now we'll try to maintain the status quo, though," Anne says. "Making a living is important, but so is spending time with the children. A farm is a wonderful place to grow up, and we want them to be able to make the most of it."

Below: Cattle grazing a herbal ley at Silkenworthy. *(Anne May)*

Tara and Roscoe Wright
Castle Hill Farm, Withiel Florey

Above: Tara and Roscoe.

Roscoe was born in 1954, in the same farmhouse where he lives now. He is a third generation farmer at Castle Hill Farm, and his daughter Tara is fourth generation.

"My grandma struggled to keep the farm going during the Second World War, but she was helped by two German nationals who had been stranded in England when war broke out," Roscoe says. "My father, Desmond, finally came back from Japan in 1947, having witnessed the horror of the aftermath of the atomic bombs."

Desmond was very keen on Closewool sheep, and he also started a fish farm in 1959, when there were only about 30 in the country.

"We used to have sixteen trout ponds, but now there are only two," Roscoe remembers. "The drought year of 1976 polished off a lot of stock, and since then the water levels have become far less stable due to changes in rainfall patterns. Predators are also a problem, especially otters and herons, so now we've got netting and mains electric fencing around the ponds."

In all, he and Tara contract rear about 16,000 rainbow trout a year for Roadwater and Taw Fisheries.

Roscoe has always been keen on plants, birds and wildlife in general – a love he's passed on to Tara. She is a bird ringer – a skill she learned while studying for a degree in Zoology at Bangor University.

"I've never enjoyed paperwork; I much prefer working outdoors with animals," says Tara. Roscoe is gradually handing over the management of the farm to her, and she has already made some significant changes. Following

Below: Tara is filling a belt feed box, which is used for very young fish. The box is on a timer and slowly releases the feed over the course of the day.

Above: Roscoe and Tara with their sheep. The different raddle colours on the ewes show when they were mated, and therefore when they are likely to lamb.

Above: Tara with some of her cattle.

advice from Chris Clark of Nethergill Associates, she has cut down the number of sheep they keep on their 220 acres (plus 40 acres of rented ground) from around 500 ewes to 300 and has replaced an enterprise finishing Charolais cross store cattle with 20 Dexter suckler cows plus followers, to decrease the need for inputs like bought-in feed.

Tara smiles. "Dad said I could have any breed but Dexters, but he's been very good about it."

"They're easier to handle than I thought they'd be, and the beef we get from them is excellent," Roscoe admits.

The next phase of Tara's plan is to sell beef and lamb direct from the farm. "It's been a huge learning curve to have sucklers and sell meat, but everyone has been so supportive," she says. "So far, I've done three beef animals, and I've just bought a second-hand cold store. From here I can either use Combe Martin Meats or Haymans at Ottery St Mary. The Meat Men, from Taunton, will pick up from either abattoir. They're excellent butchers."

The cows calve outside – mostly in May, after lambing has finished. The sheep, which are half Devon Closewools and half crosses, lamb inside by night and outside by day. The ewes are given soya-free cake prior to lambing. However, the cattle don't get any bought-in feed, so input costs have reduced considerably. This is the Wrights' first winter without Charolais cattle.

"We haven't used artificial fertilisers for years," Roscoe comments. "Exmoor has lost more species-rich grassland than woodland, so managing our pastures to encourage biodiversity is really important."

"I want to make the most of new opportunities, but I'm terrified of seeing the farm change too much," Tara concludes. "The Hill Farming Network has been a great source of information, support and friendship… I just want to find a way of being able to stay here and make a living from the land we love while looking after nature as much as possible."

Gary and Amanda Taylor
Shoulsbarrow Farmhouse, Challacombe

"I'll give you three years – you'll never stick the weather up on that hill," a local farmer told Gary and Amanda Taylor soon after they'd moved into Shoulsbarrow Farmhouse.

"We're still here, ten years later," Gary says with a smile. "Challacombe is very accepting of incomers."

Amanda admits both they and their animals had to get used to the amount of wind and rain. "When we were living in Wiltshire we rented fifty acres and had demanding full-time jobs," she says. "Gary was a furniture restorer, and I was a deputy-head teacher. We looked after our animals before and after work, so in the winter we did most of our farming in the dark. This is a doddle in comparison."

A holiday with their horses at Little Brendon Hill Farm, near Wheddon Cross, in 1993 introduced the Taylors to Exmoor, and straight away they decided they would move to Exmoor one day. They looked at many properties

Above: Amanda brushing Florence the Suffolk Punch mare. Florence is 15 years old.

Below: Gary in conversation with Betsy the Herdwick sheep. She is eight years old, and she's one of the first Herdwicks the Taylors bought.

before deciding on Shoulsbarrow Farmhouse with its 20 acres of land, and moved there with their animals in 2013.

"We started farming rare and traditional breeds of livestock in 1995," Gary explains. "We've never liked the way supermarkets treat meat, so decided to grow our own. We produce chicken, goat, lamb, beef, pork, bacon, sausages and eggs, with the occasional goose or turkey for Christmas. Our sheep, cattle and pigs go to Combe Martin abattoir and then John May does the butchering for us. He's excellent."

"Our bed-and-breakfast guests love eating home-produced food," Amanda adds. "The Black Venus at Challacombe and the Exmoor Forest Inn at Simonsbath have taken our meat in the past, but the trouble is we can't give them a continuous supply because we haven't got enough animals."

Gary and Amanda keep Shetland, Herdwick and Llanwenog sheep, Gloucestershire Old Spot pigs, "pretty, medium-sized breeds of cattle", three Exmoor ponies, a Suffolk Punch mare, Golden Guernsey and English goats, geese, ducks, chickens and a turkey called Twizzler who's now eight years old.

They've done a lot of work on the house and the land since moving, including building work, plastering, decorating, making furniture and creating soft furnishings.

"We haven't got a telly, so we can work into the night," says Gary.

On the farm, they've done fencing and hedgelaying, and they've recently planted about 100 metres of new hedgerow. Amanda shears the sheep, and they do all the farm jobs themselves. In addition to their own land, they graze another seven acres belonging to a neighbour.

Having both been salaried, one of the things they found difficult at first was charging people who stayed with them. "Being self-employed requires a different mindset," Gary says.

The Taylors took part in the SFI test and trial because they wanted small farms to be represented. "The paperwork involved with farming never ceases to amaze me," Amanda comments. "It's rather like preparing for an Ofsted inspection!"

It's clear the Taylors have stuck the weather up on the hill and are here to stay.

"In Trowbridge we lived next to a main road and opposite a supermarket," Amanda says. "We still have to pinch ourselves to check we're not dreaming."

Above: Madge is a four-year-old Gloucester Old Spot sow.

Above: Twizzler is probably the luckiest Bronze turkey in the world. He has lived through nine Christmases.

Above: Ichabod, a three-year-old English billy goat.

Chris Lerwill and Kingsley Nicholas
Combe Martin Meats

Chris and his family farm about 1,100 acres near Combe Martin, spread across three holdings: Verwill, Holdstone and Waytown. In 2008, he opened a farm shop at Waytown and, a few years later, moved into marketing his meat wholesale.

The abattoir just down the road at Combe Martin provided Chris with a crucial service, so it was very worrying when it looked as if it would close down if a buyer couldn't be found.

Kingsley, a friend and neighbouring farmer from Girt Down Farm, shared Chris' concern because he, too, used the abattoir for his home-produced meat. Buying the business wasn't a commitment Kingsley wanted to take on by himself, so he asked Chris to join him.

In February 2017, Combe Martin Meats Limited was incorporated, with Chris and Kingsley as Directors, and – after a great deal of hard work – the renovated facility opened in June that year.

Now, seven years on, the business is in good shape despite many challenges.

Chris admits things could have been different if it weren't for the unfailing support of local catering and foodservice

Below: Owners and staff in the cold store at Combe Martin Meats. (Kirsty, Chris L, Zeynep, James, Chris W, Steve, Reece, Stewart, Luke, Kingsley and John.) Beef carcasses are hung here for up to four weeks to dry-age, which reduces the moisture content and improves the flavour and tenderness of the meat.

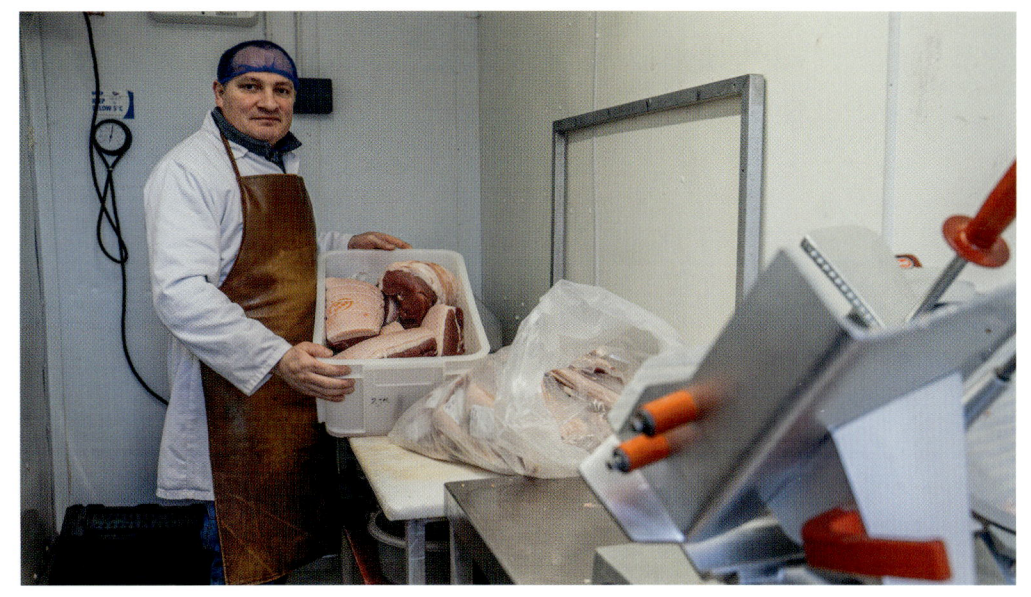

Above: John Myrda, the butcher.
Left: Pig carcasses, inspected and labelled. Pork and lamb carcasses are only hung for a few days before being cut and packed by the butcher.

wholesaler Philip Dennis, which has given them backing and guidance.

He supplies Philip Dennis with around 320 carcasses of native-bred beef a year, which go into the company's Exmoor National Park Beef range. "I sell exclusively through Philip Dennis," he says. "Provenance, welfare and food miles are really important."

Kingsley explains how his own meat business has evolved: "At first, I dabbled in selling to pubs and direct to the public, but selling individual cuts takes time so now I supply lamb, beef and pork to local butchers. I also sell lamb to Philip Dennis for its Exmoor range. In all, I put nearly three thousand lambs a year through Combe Martin Meats."

Both men agree that the support of local producers has been crucial, especially during the Covid epidemic when hotels, pubs and restaurants closed.

"A big part of Combe Martin Meats is the service it supplies to local farmers," says Kingsley. "I didn't expect there to be such a growth in that side of things, but I think buying local during Covid helped, together with increasing awareness of food miles and animal welfare."

The abattoir is open twice a week, on a Tuesday and Friday, and the staff there have gained a reputation for being skilful and compassionate. In no small part, that's due to Stuart Patterson who, in his mid-60s, is the unofficial patriarch. Another member of staff with a lifetime of experience is 70-year-old Steve Dunn. Then (in age order) there's John the butcher, Chris Lerwill, Kingsley, Jimmy, Luke, Kirsty, Chris and Rees. A vet is always in attendance as well.

"Everyone does a fantastic job," Chris says. "They start at five-thirty in the morning and work through until about three-thirty in the afternoon – or later if we're busy."

Up to about 25 cattle, 60 pigs and 200 lambs are killed there every week. Around 50 goats are killed each year as well.

Combe Martin Meats must be the only abattoir in the country that's situated in the middle of a village. "It isn't ideal," Chris admits. "The plan is to move to Waytown, where there's more space and better access, but we've got to do things in the right order so there's minimal disruption. Inflation, increased building costs and high interest rates are all challenging at the moment, but we'll get there. What we don't want to do is put anything at risk through debt."

Kingsley agrees. Three years ago he bought Fullaford Farm, near Brayford, where he lives with his partner and their young daughter Farley. He's farming 300 acres.

"It's a juggling act," he says. "The worst thing is making sure you're not neglecting anything, but I'm glad we've taken the opportunities as they've come along."

Christmas Fatstock Show

Anti-clockwise from the top: The judges have made their decisions; after the judging, everyone can take a look at the results; Ron Delve looking at the turkeys; Mike, Sue and Charles Hayes' champion Texel x lamb. *(Emily Fleur)*

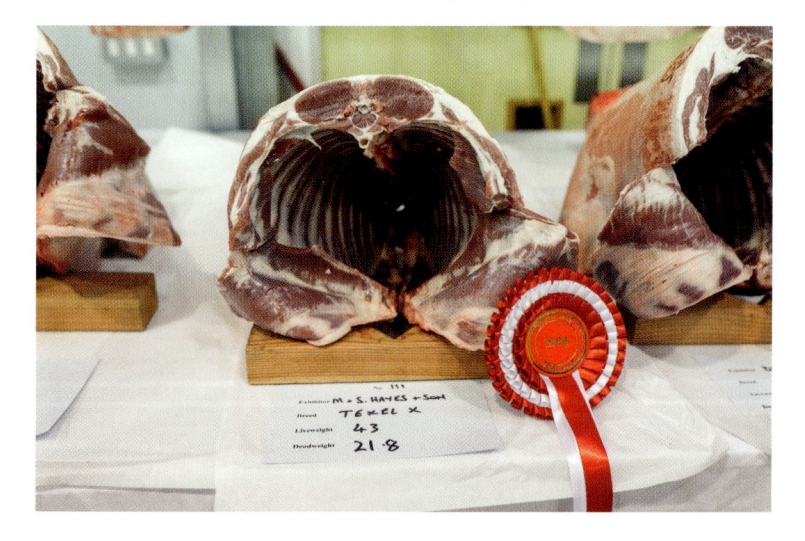

160

Brian and Clare Westcott
Beech Tree Farm, Exford

Brian's great-grandfather was a Cornish miner who came to Exmoor in the nineteenth century to manage Newland Quarry (between Simonsbath and Exford) for the Knights.

"It was the only place where lime was produced on Exmoor – you can still see the pits and lime kilns there," Brian explains. "Grandfather built the bungalow where my Mum lives at West Ley Farm in the early 1950s, when there was a seventy-five per cent grant for building new houses on Exmoor. Clare and I now live on the opposite side of the road at Beech Tree Farmhouse, which we built in 2007 when attitudes towards building houses within the National Park were rather different."

Clare was brought up at Tidicombe Farm, near Arlington, where her dad farmed, and her mum had Arlington Riding Centre. "I've always loved farming – I especially love cows and calves," she says.

"I suppose you could say I landed in farming," Brian says. "As a schoolboy at Brymore my eyesight got progressively worse, and at sixteen I was diagnosed with macular degeneration, which affected my ability to drive and the jobs I could do. I learned hedging and shearing at Exmoor Young Farmers, and did both to earn money. Dad also taught me a lot about hedging, and we used to hedge together. The thing I found most difficult when Dad died was going back on a hedge again."

Brian and Clare started off with 68 acres in 1992. They've bought 56 acres since then, and they rent 120 acres. To begin with they had Closewools, but they replaced them with Poll Dorsets, which will lamb at any time of year. Winter lambing their older ewes means they can sell them as couples in Exeter Market when prices are at their highest.

Their Hereford cross Friesian cows are put to an Angus bull for calving in May and June. The youngstock are sold at 12-18 months old at Cutcombe.

Mark, their elder son, was the first person from either family to go to university. He graduated with first-class

Below: Brian and Clare with some of their Poll Dorset cross lambs, which are grazing on a herbal ley pasture.

Above: Brian checking his cows and calves. *(VE)*
Below: Brian and Clare enjoying a day out at Cutcombe Fatstock Show. *(Emily Fleur)*

honours from Exeter, and now works as a design engineer for Babcocks in Plymouth. He lives in Alcombe, near Minehead, with his wife Bex and children Melanie and Reggie.

Clare and Brian's younger son Andrew lives at home, works in construction and helps out on the farm with jobs like hedging when he can.

"I don't think people appreciate small family farms," Clare says. "There's a huge connection to the land and the community, with knowledge passed down through the generations."

Brian has been involved with the EHFN from its launch in 2014. He is a Director, and both he and Clare are keen participants.

"It's lovely to be with like-minded people who understand, and I'd never have gone to Northern Ireland without the Women in Farming Group," says Clare. "There are good people at the core of the Network – everyone is so supportive."

Brian agrees. "And the Farm Resilience Programme that Charmain Dascombe organised a few years ago was good, too," he adds. "On our farm we're constantly assessing what's working and what isn't. We love our home, and we've got to adapt to stay here."

The Fatstock Show at Cutcombe marks the beginning of Christmas for the Westcotts. The camaraderie amongst competitors is wonderful, they say, and it's very interesting to see the differences between the live judging and the judging of the carcasses when they come back from Stillmans Butchers three days later. This year they didn't win a prize, but they still had a great time.

The Anstey YFC Tractor Run

Over 100 tractors took part in the Christmas Tractor Run, which started near Oldways End and travelled through Exebridge, Bridgetown, Exton, Winsford, Withypool, Molland Moor, West Anstey and Yeo Mill. The event raised £2,293 for the Devon Air Ambulance and Exmoor Search and Rescue.

Far left: Paul and Reuben Gillbard's tractor from Hele Barton Farm, near Black Dog.
Left: George and Kieran Vigars from Exford.
Below: The winning tractor of Martin Thorne from Crossways Farm, Bratton Fleming.

Charmain Dascombe
Farming Community Network Support Officer

Charmain grew up at West Luccombe with her younger brothers Geoff, Dick and Robin. Her father John Tucker and mother Rosemary (née Baker) farmed about 1,000 acres at Lucott Farm, West Luccombe and West Lynch. "I loved the farm animals – it was my job to nurse the poorly ones," she says. "And I was pony-mad. Dad and I used to go riding and hunting together. He taught me a lot, and so did my great-grandfather Bill Harding."

Always a team player, Charmain was a keen member of the Devon and Somerset Pony Club and Cutcombe Young Farmers, and she became secretary of both Cutcombe and the West Area Group of Young Farmers. Her first job was with Shearwell at Wheddon Cross, after which she moved into veterinary work, qualifying as a veterinary nurse at Langford in Bristol and becoming the first qualified veterinary nurse at White Lodge Vets, Minehead.

In 1992, when she was 20 years old, Charmain married David Dascombe. Their son Oliver was born in 1999, followed by Rebecca in 2001 and Archie in 2003. During that time, she carried on working part time at White Lodge.

Somehow, she has also found the time to be Secretary of Porlock Show since 2000. "It's something that's very close to my heart," she says. "Grandad started the show to raise money for Porlock." The committee has worked hard to keep a traditional local show, with sheep classes and stags' horns competitions as well as horse and pony classes.

"Mum died from breast cancer when she was 50 and I was expecting Archie," Charmain remembers. "It was a turbulent few years, what with that and the foot and mouth crisis."

Farmers were being encouraged to diversify, so Charmain opened a cattery a week after Archie was born. It meant she could be at home with the children, and she carried it on until they were all teenagers.

With encouragement from her family, she became a lecturer in veterinary care at Bridgwater College in 2015. Soon after that she also took on running the Farm Resilience Programme delivered on behalf of The Royal Countryside Fund, which reminded her how much she enjoyed working with farmers, and just before the Covid epidemic she became an online teacher for the College of Animal Welfare.

Two years ago, she started working for the Farming Community Network (FCN), covering Somerset, Dorset and Wiltshire. "I work with an amazing group of people – most of them volunteers. We're all there because we want to help, and there's a lot to do because there's so much stress and uncertainty in farming at the moment," she says. "Finances, the phasing out of the Basic Payment Scheme [BPS] and succession are major problems, along with stress due to TB and farm inspections. I feel honoured that

Below: Charmain with Jude McCann, CEO of the FCN, at the Royal Cornwall Show 2023. *(Emily Fleur)*

people will speak to me about very personal things, safe in the knowledge that everything is completely confidential. We're often told we've really made a difference, and that makes it all worthwhile."

Charmain is also a Trustee of the Exmoor Rural Health Hub, which she helped to set up at Cutcombe Market, and it was her idea to have a carol concert (which became known as Singing in the Ring) in the auction ring. "David Weir was incredibly enthusiastic – so was Katherine. We wanted everyone to feel included. My work with the FCN has shown me that the farming community on Exmoor is really special. Farmers in other areas haven't got what we've got here; it's something to be cherished."

Very sadly, Charmain didn't feel well enough to attend Singing in the Ring because she had just started treatment for breast cancer. "I have a scan every year because of what happened to Mum, so in a way that gave me a false sense of security. It was only by chance that I checked myself and found a lump. It's quite an aggressive form of cancer, so the hospital is treating it very intensively," she says. "It's ever so scary, but everyone's been incredibly kind and supportive. It's hard to put my life on hold when I really enjoy my jobs and the farm doesn't stop around me. At least the children are grown up now, and Oliver's working with David on the farm. I've got a long way to go with the treatment, but I'm determined to be at the next Singing in the Ring."

Above: Charmain at the Balmoral Show with the Women in Farming Group. L - R: Alison Habberfield, Charmain Dascombe, Angela Andrews and Fran Willes. *(EHFN)*

Left: Charmain at the Devon Closewool Centenary event, talking to Pat Williams. *(VE)*

Singing in the Ring
Cutcombe Market

A week before Christmas, Exmoor's farming community gathered at Cutcombe Market for 'Singing in the Ring', a carol service organised by the EHFN, the FCN and Exmoor Churches, with supper provided afterwards by Heywood's Catering.

The Royal Countryside Fund (formerly the Prince's Countryside Fund) contributed towards the event.

Katherine Williams, Charmain Dascombe, David Weir and Lucy Heywood had worked hard for several months to turn an idea into reality, and many others helped – Adam and Oliver Hill provided small bales for the event, for instance, and Thorne's Butchers provided beef for the supper. The auction ring was festooned with Christmas lights, over £600 was raised for the FCN and Exmoor Benefice, and the whole occasion was a wonderfully inspiring way to round off the year.

The service was conducted by Rev. David Weir, and it brought together over 230 people of all ages from across Exmoor, including children from Cutcombe and Exford First Schools. They led the first carol, *Once in Royal David's City*, helped by the Exmoor Benefice Choir, with Brenda Staples playing the keyboards and Bernie Bettinson playing his cornet. The carols that followed were also sung with great gusto. Peter Huntley did a reading, and prayers were said by Lucy Heywood for the Market, Keith Branfield for the Café, Kim Dart for the EHFN, Olivia Winterton for the Health Hub, and Ian May for the wider community. Andy Jerrard introduced and led The Lord's Prayer.

Above: David Weir welcoming everyone to Singing in the Ring. *(Emily Fleur)*
Below: The choir and musicians in the auction ring. *(Emily Fleur)*

Above: Singing in the Ring was a wonderful event that brought people together from all over Exmoor. *(Emily Fleur)*

As Charmain was too ill to take part in the service, Katherine read out her address, which stressed the importance of the market as a community hub.

"The community across Exmoor is more than a collection of people doing their own thing on their own land. It is a community of people who know each other and look out for each other," Charmain's message said. "Today we gather to celebrate our community… The Market, the Health Hub, the Café, the Hill Farming Network, the Farming Community Network, the market chaplains and vicars are all here to remind us of how much we are loved and valued, regardless of the challenges and how well we feel things are going; to hold out a hand of welcome and encouragement, especially when we might be feeling alone or afraid."

Everyone's thoughts were with Charmain as her message was read. Then Katherine added a few words of her own, thanking the people who had worked so hard to make the evening such a success, especially David Weir. "David, thank you for bringing Christian values from the church to the market throughout the year," she said. "You are the pillar of our community, having a gentle, unassuming presence and being so supportive across Exmoor, both to individuals and the community as a whole."

Above: Keith Branfield talking to Robert Nancekivell. *(Emily Fleur)*

Afterwards, somebody commented to David Weir that the problem was if Singing in the Ring became an annual event, it would soon become an unmissable occasion in the Exmoor calendar, with more and more people turning up.

He smiled and said, "That won't be a problem."

January

The New Year is often a time for assessing how things have gone during the previous year, identifying where changes should be made and planning how to implement them.

On Exmoor, new challenges and opportunities due to changes in government support for agriculture and trade deals following Brexit, plus global problems like climate change, are prompting farmers to rethink how they manage their farms.

Like Andrew Pugsley, many farmers who used to focus on the production of livestock are now looking to cut inputs and maximise their income from the land, using livestock as a management tool.

This has prompted renewed interest in native breeds and their value as conservation grazers, together with regenerative farming methods. However, a rise in the tick population is forcing some farmers, like Richard Langdon and Tom Burge, to rethink where and when they graze their livestock.

As dairy farmers, the Nicholls family must produce a large amount of top-quality forage for their high-yielding cows. Cutting inputs is much more difficult on a dairy farm, but they are keeping fewer cows than they used to.

The Molland Estate
Graze the Moor Project
Christina Williams and Richard Langdon

Above: Christina Williams on Molland Moor.

The Molland Estate has 10 tenanted farms plus 1,680 acres of moor, and the land has been in Christina Williams' family for hundreds of years. Molland Moor is not a common but has always been managed by annual licences. Up to 15 farmers used to stock it with cattle, sheep and ponies. However, the traditional management of the moor changed when winter grazing by cattle and sheep was stopped, fewer livestock were allowed on it during the rest of the year and swaling was restricted. The shift from headage payments to land-based payments also encouraged farmers to replace hardy livestock with fewer, more productive cattle and sheep. Understandably, they were reluctant to risk grazing the moor with these more valuable animals, especially as TB and tick-borne diseases were on the rise.

"The area is a Site of Special Scientific Interest because of its heather moorland, but the heather was declining at an alarming rate as invasive species like *Molinia*, gorse and bracken took over," Christina explains. "So I approached Natural England to set up a moorland management experiment that included winter grazing, swaling, spraying herbicides, cutting and re-seeding."

The Graze the Moor project began in 2010, supported by the ENPA Partnership Fund and, more recently, a FiPL grant. The arrival of Richard Langdon and his father Steve was key to delivering the project. Richard agreed to stock the moor with sheep and cattle from Luckworthy Farm as part of his tenancy agreement.

"I was seventeen at the time, so I'll have been here eleven years this spring," Richard says. "I love farming on Exmoor

area-based payments," Andrew says. "We used to have about fourteen hundred ewes and a hundred cows; now we've got four hundred and fifty ewes and a hundred cows.

The cows are mainly Angus cross Simmental or Limousin, and they are put to a Charolais bull to produce commercial calves for the suckled calf sales at Cutcombe Market. They graze extensively on about 1,800 acres for most of the year, only being housed from around the beginning of January until after calving in March and April.

Andrew has used cut and baled *Molinia* and rush for bedding since 1998. The compost from them contains higher levels of phosphate and potassium than straw, and both Andrew and his vet believe that exposure to ticks through bedding as well as moorland grazing gives the cattle increased immunity against tick-borne diseases.

"Following the foot and mouth year of two-thousand-and-one, coupled with the depressing livestock sales of that period, I started thinking about moorland restoration as

Above: Sheep in the shed at Warren Farm.
Below: Warren Farm.

a way of introducing additional income for the business," Andrew remembers. "The Exmoor National Park Authority, Defra and English Nature (as it was then) all became involved in a five-year plan that we called the

Warren Moorland Restoration Plan. Our aim was to reverse the spread of *Molinia* and encourage a balanced mixture of different moorland plants."

Geoff Eyre, a contractor in the Peak District, was restoring heather moorland there with great success. He pioneered methods of collecting and cultivating the seeds of many different moorland plants, including heather. Knowing that fire stimulates heather germination, he developed a method of collecting a smoked solution based on heather burning and treating the heather seeds he had harvested with it, which resulted in a marked increase in seed germination.

Andrew studied Geoff's methods, and set about building his own heather harvester that sat on the quad bike. He also built a burner and condenser so he could marinate the harvested seed to optimise germination rates. Ground to be seeded was sprayed with glyphosate, the dead vegetation was cut and baled for bedding, and seed was applied by a number of methods – including broadcasting with an ATV spinner and via a small slurry spreader.

"Heather seeds have the consistency of ground cinnamon – they're tiny," he says. "If you re-seed the way Geoff does it, with a purpose-made twenty metre fan sprayer, you only need a handful to re-seed about eight acres."

Baling heather and rolling it out didn't work nearly so well; neither did sowing seeds without removing the existing vegetation. "If you cut *Molinia* and don't bale it, it'll come back even thicker – it's got a life of its own," Andrew says. "We've found that burning is key. The heather on Mill Hill is recognised as the best on Exmoor, simply because the Woollacott family have maintained the traditional method of rotational burning."

Heather re-seeding on the trial plots worked well, but it was expensive. Since the project ended 15 years ago, no more heather has been introduced. However, Andrew is still grazing, burning, cutting and baling to control the *Molinia* and encourage the introduced heather and previously smothered species like whortleberry, heath bedstraw, purple heath orchid and bog asphodel.

Andrew and Linda Pugsley
Zeal Farm, Hawkridge

Above: Andrew with his dogs: Ted, Chip (the tip of his ear is visible under Andrew's arm) and Spud.

"You've got to farm with nature up here. You soon find out if you've pushed things too far," Andrew Pugsley says.

He and his wife Linda live at Zeal Farm, near Hawkridge, and Andrew is a Director of the EHFN. Linda provides bed and breakfast in the farmhouse. She ran a catering business for 30 years, but is trying to scale that back now – despite still being in demand for special occasions like the Closewool Centenary celebration at West Whitefield last August. In all, the Pugsleys farm about 1,350 acres; 850 acres of that being moorland and rough grazing but not common land. They used to have 300 suckler cows and 1,500 breeding ewes, but now they farm about 150 suckler cows and 700 breeding ewes, so they've pretty well halved their livestock numbers.

The cows are Hereford or Angus cross Simmental, which are put to a Charolais bull to produce strong stores that are sold, TB tests permitting, at 12 to 15 months old. The ewes are also commercial crossbreds: North Country and Suffolk Mules that scan at around 170 per cent. Both the cattle and the sheep suit a system with fairly high inputs. The cattle are in-wintered but calve outside in April (which wasn't easy in the bad weather last year), and after weaning in the autumn the calves are kept in until they're sold. The sheep remain outside for most of the year, but are housed in a shed for lambing from the end of March for a month. After lambing has finished, the ewes and lambs go to grass keep so that the fields around the farm can be laid up to grow grass for hay and silage. It's a system that used to work well, but Andrew is now having a rethink.

"I'm grazing lighter and taking one crop of hay or silage off the land rather than two," he says. "Since fertiliser reached eight hundred and ninety pounds a tonne a couple of years ago, I've hardly been using any. I'd rather take extra grass keep than buy fertiliser, but everyone is now going into Countryside Stewardship so grazing restrictions are much tighter. Another driver for change is TB, which has been a major issue for the past five years. We're on testing every sixty days at the moment, and it has completely upset our marketing."

Andrew is exploring the possibility of switching to native breeds and keeping them out all year round. To begin with he's looking for 50 Devon cows to keep on the moorland around the house. He and his brother Steven grew up at Zeal when their father John was farming it, and he says old photos show much more heather on the moorland than there is now. He'd like to get it back to how it was, and that will involve grazing. "Ticks could prove to be a problem," he says. "We've definitely got more than we ever used to have."

However, he hopes that by treating the cattle for ticks

Above: Andrew and Linda.
Below: Cattle grazing during the summer on Old Barrow Down (otherwise known as Lords), with Molland Moor in the background.

every couple of months when they're in for TB testing he'll be able to minimise the risk of tick-borne diseases.

"We used to try to maximise our income from the livestock, but now livestock are becoming management tools that help farmers to fulfil the ELM agreements on their land," Andrew concludes. "Our focus has shifted to maximising our income from the land itself rather than the livestock we can rear on it."

Tom Burge
Oaremead Farm, Malmsmead

Some people may think that winter grazing doesn't need to be managed because the grass isn't growing much, but Tom Burge would disagree.

"I work out my rotational winter grazing plan when I make hay," he says. "All the hay bales are stored outside in the fields where they've been made, and the fields aren't used again until they're needed for bale-grazing in winter. Spoilt hay around the outside gets trodden in to feed the soil, so nothing's wasted."

Tom uses electric fencing to keep his suckler cows in paddocks of about an acre per day when they're bale-grazing in the winter. "The thickness of the deferred sward helps to hold the cows up so they don't damage the soil, but if they start making a mess I'll move them on," he says. "I'm learning how each field handles differently, and I'm constantly having to adapt to the weather."

Before settling on a farming career he trained as an engineer, and he enjoys creating or adapting machinery. A good example is the Burge Bale Unroller (BBU): a galvanised bale unroller that can be hitched onto an ATV or UTV, minimising soil compaction while saving time and energy when feeding round bales to outwintered livestock. Demand from all over the UK has increased at such a rate that PFG Fabrication from Bampton makes BBUs for him now.

"It's a whole new skillset, selling to the public," he comments. "This year I'll be attending Groundswell, for other like-minded farmers to see it." He's fascinated by regenerative farming, and is an enthusiastic member of the Precision Grazing discussion group.

Above: Tom moving electric fencing to give his livestock a fresh area to graze.
Left: Tom's Scotch Blackface sheep do well on moorland.

Recently, he has replaced his Charolais bulls with Aberdeen Angus bulls for his pure Angus cows, aiming for smaller cattle that are less damaging to the ground and more carbon efficient. This spring he'll be outdoor calving for the first time from mid-April, to coincide with spring grass growth.

Cow numbers have increased from 100 to 130 this year. He'd like them to go up to 150 - 180 because he reckons that'll be the 'sweet spot' for his farm. He's finding he can make more profit from sucklers than from sheep when they're kept outside on a low input grass-based system.

His goal is to produce zero-carbon beef, and in preparation for this, and maybe also for carbon trading, he has done a carbon soil audit of the whole farm.

"Long-term, I'd like to sell my own beef and mutton," he says. "In fact, I'd love to have a farm shop and café in Malmsmead."

Last autumn, he cut down his commercial ewe flock from 1,100 to 650 Romney cross ewes, primarily because of tick-borne diseases. The main problem is louping ill. It's particularly bad on ground frequented by deer, so he's adjusting his grazing management to reduce the risk.

The ewes will lamb outside from the middle of April, which is also when the cows will be calving.

In addition, 300 purebred Scotch Blackface ewes are kept out on the moor, and the wethers graze Brendon Common until they're three years old. Tom is building a market for Exmoor mutton, which he'd like to get into local pubs and restaurants.

With such volatile prices for commodities, Tom is looking to maximise environmental payments on his farm. "I haven't spread any fertiliser or ploughed any land for a few years now, but I'm growing more grass than ever, and the increase in both biodiversity and the amount of clover is amazing," he says enthusiastically. "I'm realising farming is about looking after the livestock above *and* below ground. The whole regenerative thing has made livestock farming exciting again."

Tom is a Director of both the EHFN and EFLA. His wife Sarah works part time for Shearwell and on the farm. Their daughter Edith will start school in Porlock in September.

Above: Josh Collins, who works for Tom, using the Burge Bale Unroller.
Below: Outwintering Angus cattle on hay at Oaremead Farm.

View from Ridge Road near Anstey Common.

James Gregory
Lower Ley Farm, Luxborough

James Gregory started scanning sheep seven years ago. "I became interested in the job after seeing Dad's flock scanned," he says. "So when I was seventeen I took the plunge and purchased a scanning machine. A two-day course was provided in Scotland, and that was it – away we went to try and build our own businesses."

When his dad sold his livestock in 2019, James bought a few of the ewes. Since then, he has built his livestock numbers up, keeping them on the 24 acres at Lower Ley Farm, which his dad still owns, and renting 90 acres of land from a neighbour.

He now runs approximately 200 ewes and followers (mostly Texel and Suffolk Mules, but also 20 pedigree Blue Texel ewes) and 20 head of cattle (Limousin cross and British Blue cross cows).

His Blue Texel sheep have won many prizes. "I got into showing after I purchased some pedigree sheep from my girlfriend's dad," he says, adding that it's a good shop window and a great way of meeting fellow farmers.

"I think scanning is majorly important for successful farming, especially with the rising cost of inputs," James comments. "It means farmers can sell barren ewes and cows sooner and feed their stock efficiently. Knowing whether ewes are carrying singles, doubles or triplets, for instance, means you can feed them the right amount and manage them when they give birth – if possible taking one lamb away from a triplet and fostering it on to a ewe giving birth to a single."

James started scanning cattle four years ago, following training from a vet. "It's a very different skill – sheep being scanned externally and cattle internally – but most of my customers have sheep and cattle, so it made sense. The sheep-scanning period is from October to late February, whereas the cattle are usually August to November. My stock are housed in the winter, so I can look after them before and after work. Luckily I have good lights in the shed!"

Right: James can tell, from the image on the screen, whether a ewe is pregnant and, if she is, how many lambs she is likely to have.

Top: James operating his sheep scanning equipment. Because scanning mainly takes place in winter, it can be a cold job.

Left: A ewe being scanned.

John Tucker
Stetfold Rocks, Exford

John Tucker's grandfather was born in 1847, and he was born in 1947. "I'm the fourth generation here," he says. "But they've been long generations!"

In all he farms about 1,400 acres, which includes 400 acres of moorland grazing at Simonsbath around the Fortescue Memorial, on a 40-year lease and farm business tenancy with the ENPA. "We used to graze hundreds of sheep until mid-winter, but now we just put our autumn calving cows there, and the ewes after weaning in August," John remarks. "There's more heather there now; also more ticks – and more deer."

The cows on the farm are Devons and Hereford or Angus crosses, and the sheep are mainly Exmoor Horns and Exmoor Mules. John bought some Swaledales a couple of years ago, but he says it just confirmed you can't really beat an Exmoor Mule. He is now planning to reduce his livestock numbers and go back to Devon cattle plus Exmoor Horn and Exmoor Mule sheep.

Since he took over the farm in 1963, John has employed numerous young adults, often giving them their first jobs or an apprenticeship placing. At the moment, Adam Hill from Timberscombe, Emily Chesterton from Porlock and William Everett from Exford are working for him on different days. "They're all good, experienced workers who use their own initiative and get jobs done without having to be asked."

Jack Allen from Bridgetown is due to start an apprenticeship when the school year finishes. "He's not from a farming background, which can be an advantage as I can teach him from scratch," John comments. "A major problem for apprentices is transport, until they pass their driving test."

His wife Kathryn works with students at West Somerset College. "Many have mentioned how much they would like to work on a farm," she says. "It's a shame the Agricultural Unit there closed down."

Seeing the people who have worked at Stetfold Rocks doing well is very rewarding.

"Maddie Norman was here for four years – she's now working for the vets," John says. "Then there's Ricky and Gary Atkins, Robert Branfield, Duncan Westcott and Fred Jones, to name a few – all hardworking young people with good futures ahead of them."

Below: John in the farmyard at Stetfold Rocks.

Geoffrey Illing
Farming Apprentice

"I've been interested in farming my whole life," says 16-year-old Geoffrey Illing. "I love it all. I especially love sitting in tractors, but it's the animals that earn the money, not the tractors."

Geoffrey lives at Honeymead, near Simonsbath, where his mum Karen Stevens is the groom and his father Graham Illing is a fencing and hedging contractor. "My sister Emily likes the countryside and farming, too, but she's mainly into horses," Geoffrey says. "My auntie Jenny Stevens farms at Codsend, near Wheddon Cross, and I've always enjoyed helping out there. She keeps Devons and Exmoor Horns. Also, I've been in Exmoor Young Farmers for four years."

Because he was born in August, Geoffrey had taken his GCSEs by the age of 15. "I enjoy the practical side of things much more than sitting in a classroom, so I decided to leave school and do an apprenticeship. We know the Edwards family, and they agreed to take me on, which is ideal."

Jill Edwards used to teach Geoffrey when he was at Exford First School. "Taking on an apprentice is definitely a holistic sort of thing where everybody's involved," she comments. "He's like a member of the family, really. Niki Gibbons, his tutor, is in regular contact with us – she visits all the farm placements."

Geoffrey goes to Cannington, near Bridgwater, one day a week, and for the rest of the time he's either working at Westermill or back home. The apprenticeship takes nearly two years, and after that he'd like to go travelling. His dream is to have his own farm.

"An apprenticeship is all about finding the right person to take you on – someone who's prepared to teach you," he says. "If you're keen, and willing to put in the time, that helps a lot."

It certainly does. "If the future's in the hands of youngsters like Geoffrey, we've got a good chance," Jill says.

Above: Geoffrey in the cowshed at Westermill.

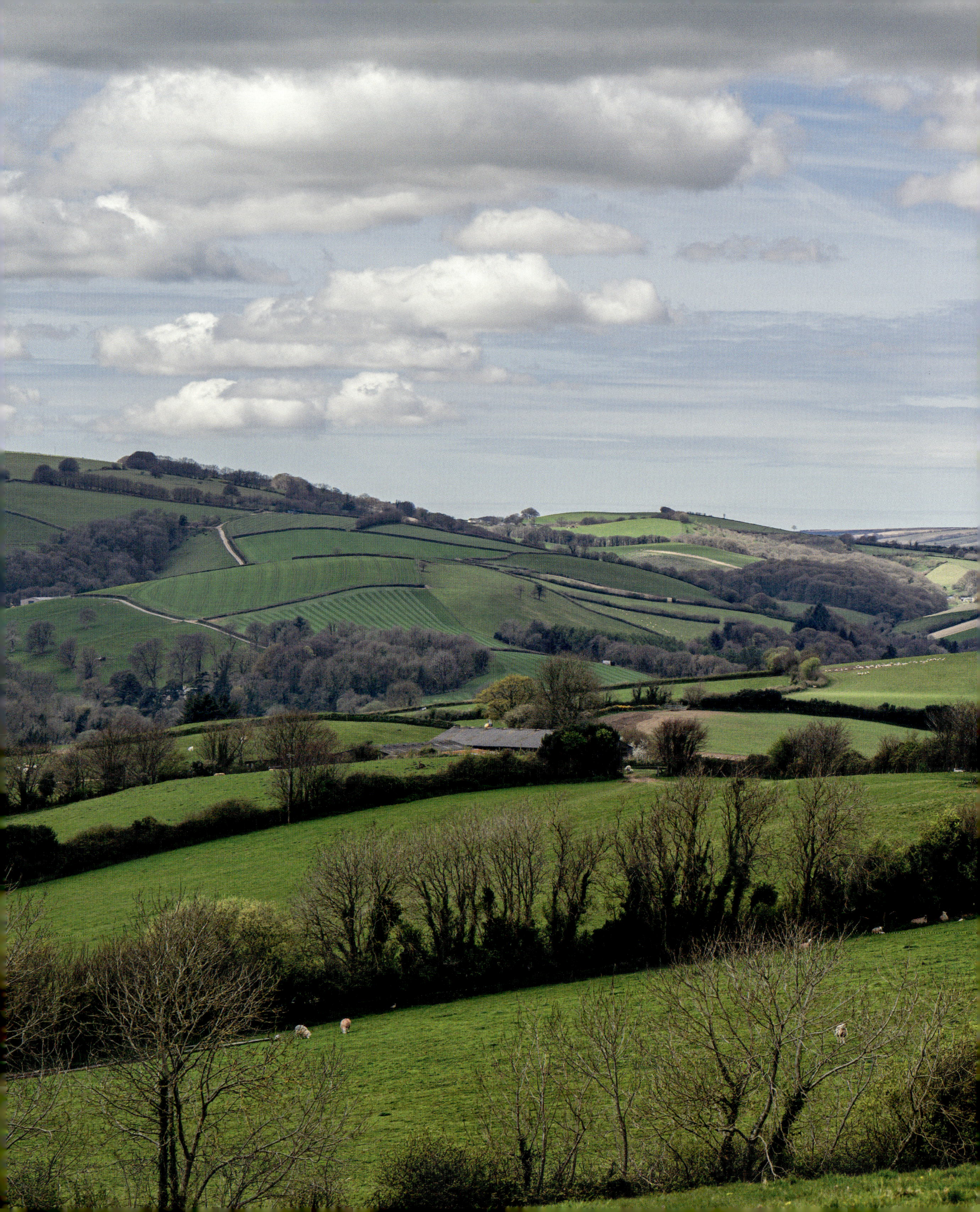

Coleridge Way at Treborough looking towards Dunkery.

Jack Bishop
Lower Court Farm, Treborough

"I never wanted to do anything else but farming," Jack says. "One of the GCSEs I studied at West Somerset College was Agriculture – with Bill Liversidge at the Farm Unit, which I really enjoyed – and I spent all my school holidays working at home. After leaving school, I did a few days a week for Eric Parker and Neil Gregory. Then, at about twenty-six, I started working for David and Charmain Dascombe part time and at home for the rest of the time."

When Jack was growing up, his father Richard and uncle Michael farmed in partnership together, with responsibility for the sheep and cattle respectively. Jack and his cousin-in-law Carl Hobbs now have a similar arrangement, and it works well.

Lower Court Farm is 320 acres, and roughly 300 additional acres are rented. Jack and Carl have about 600 Mule ewes and 65 Continental cross suckler cows. The calves are usually sold at the Cutcombe two-day suckled calf sale, and the whole herd is TB-tested six weeks before the sale so the calves can be creep fed if they're going. About 10 acres of barley are grown every year to feed to the cattle. Their lambs are kept for as long as possible to catch higher prices at the beginning of the year.

Recently they have used less fertiliser on the farm and have started experimenting with rotational grazing. "It's worked well, and we're keen to extend it, but there are a lot of rights of way over the farm, including the Coleridge Way," Jack says. "We don't want anyone to get electric shocks from the temporary fencing needed to divide the fields."

Jack and his wife Sally have two young boys, and Carl and his partner Ellie also have a boy, so they are farming with future generations in mind. "We're trying to make the most of new opportunities without altering the farm system too much," Jack explains. "We could jump into the new environmental schemes with both feet, but we're still young and keen to farm."

Below: Jack and Carl in one of the fields, with Watchet, Williton and the Quantocks behind them, and the Bristol Channel and Welsh coast in the distance.

Helen and Sarah Thomas
Higher Westland Farm, Blackmoor Gate

Sisters Helen and Sarah live at Higher Westland Farm, just inside the National Park boundary near Blackmoor Gate. Their parents, Hugh and Barbara Thomas, keep around 500 Lleyn ewes and also have seven home-bred Thoroughbred horses on the farm, two of which are in point-to-point training. Barbara works for Bridgmans agricultural merchants at Porte Farm, about a mile away. Both girls combine helping on the family farm with careers at the local veterinary practice.

After Helen studied Chemistry at Cardiff University, she got a job working for Torch Farm and Equine Vets at Mullacott, near Ilfracombe. Then, three years later, when Sarah graduated from Southampton University with a Chemistry degree, she joined the client-support team there.

"Jobs were available at Torch when we needed them; we fell into them and never left," says Helen. "I control the pharmacy stock part time, which is ideal because I can help with the horses or farming before and after work."

Sarah now leads the reception team. "I work more hours and help at home in my spare time, and we all love point-to-pointing at the weekends – it's always been a big part of our family life."

Hugh Thomas is Chairman of the local point-to-point committee, and Barbara is Secretary.

Barbara is also a Churchwarden of Holy Trinity, Challacombe.

Lambing is another occasion when the whole family pull together. Older sister Claire even takes time off her work as a civil engineer in Bristol so she can return home to help.

Both Helen and Sarah say Torch is a lovely company to work for, and a lot of local people with farming or equine backgrounds are employed there.

"Clients are often stressed when they ring, so it's really important for them to be able to talk with someone who understands," Sarah says.

"It makes a huge difference that it's still an independent business – so many vets are corporate nowadays," Helen adds. "We both enjoyed university, but home is where our hearts are, and working for Torch Vets has meant we can stay here."

Above: L-R: Helen with Peg and Sarah with Tess.

Shearwell Data Ltd
Putham Farm, Wheddon Cross

With over 140 staff, Shearwell Data is the largest employer within Exmoor National Park, and it has given many of Exmoor's youngsters the chance of a good job since it began in 1992 with just a few employees.

"It's important that young people can see opportunities are available in the National Park," says Richard Webber, who started the business with his wife Carolyne. "Second-generation employees are now joining Shearwell, with a number already holding management positions."

Richard initially had a business called Shearwell Shepherding, providing mobile dipping and shearing as well as selling lambing equipment. Then, when he was shearing in Norway, he purchased an ear tag making machine, imported it and started manufacturing ear tags to sell for animal identification.

The business grew rapidly, and Shearwell is now the market leader for both visual and electronic identification (EID) tags for sheep and cattle in the UK. It exports to more than 35 countries and is also manufacturing in Australia, New Zealand, Canada and the USA.

"There are different regulations in different countries," Richard explains. "And the plastics we use have to perform in a wide range of conditions, from hot deserts to cold tundra. We are producing tags for an increasing number of species, too, including deer and bison."

Besides producing tags, the company has developed complete livestock management packages with programmes, readers, software and handling systems that allow farmers to collect and organise data so they can make informed business decisions. Despite all the technological wizardry, though, good customer service is at the core of the business, which is well-known for having a competent person to talk to immediately if you telephone.

"My passion has always been livestock farming," Richard comments. "Everything we do is about providing livestock farmers with reliable, useful products and first-rate customer service."

Richard, Carolyne and children James, Emma and Sam are all Directors of Shearwell Data Ltd. The Webbers are farmers as well, keeping sheep and cattle on their farms near Wheddon Cross. The farms also provide opportunities to test new ideas, and in this constantly evolving business you can be sure there will be plenty of those.

Above: L-R: Emma, Carolyne, Richard, James and Sam. *(supplied by Shearwell Data)*

Lucy Gill
Trainee Rural Chartered Surveyor

Lucy and her brother Timmy grew up at Yealscombe, just outside of Exford, as their mum Lesley worked for Lady Heseltine as caretaker, housekeeper and groom. Their dad Andrew ('Gilly') was a self-employed farm worker. One of his workplaces was at Riphay Barton for the Yandle family, and Lucy has fond memories of helping him with things like lambing. She and Timmy were members of the Devon and Somerset Pony Club; they both loved horses, hunting and country pursuits.

After A levels, Lucy worked for Mole Valley Farmers in the procurement team, eventually becoming a buyer. However, the Covid lockdown prompted her to look for a change in direction. By chance, she was having supper with Katherine Williams when Kevin Bateman, of Bateman Hosegood Chartered Surveyors, rang to say he was looking for an administrator. "I got the job, but as well as doing admin I also shadowed Kevin doing professional work," Lucy says.

She is now three years into a five-year course at Harper Adams University, studying for a degree in Rural Estate and Land Management. This involves being at university for about nine weeks and working for Bateman Hosegood for the rest of the year.

"It is a government-funded course, with support from my employer," Lucy explains. "I'm really enjoying it, learning and putting it into practice straight away, which is key to keeping up with evolving changes. It's reflects my interests and I love being out and about helping farmers… What's happening at the moment could completely change the way we farm and manage the landscape. It's all becoming so complex, and even people who used to do all their own form-filling are now turning to professionals."

Lucy's work is predominantly in Exmoor and North Devon. "Exmoor is home for me," she says. "It's a very special place – the community, the people and the Network. I'm so thankful I got the opportunity at Bateman Hosegood, and it's really nice to be following in the footsteps of my grandad Tom Rook."

Below: Kevin Bateman with Lucy, visiting a farm.

The Farming in Protected Landscapes Team

The Farming in Protected Landscapes (FiPL) programme is a part of Defra's Agricultural Transition plan. It started in 2021 and is expected to run until March 2025, offering funding to farmers and land managers within National Parks or National Landscapes in England for one-off projects that do more for 'nature, climate, people and place'.

On Exmoor, this national programme is delivered with support and advice from the ENPA's FiPL team: Charlotte Thomas and Sarah Eveleigh (FiPL Officers), and Vickie Sellick (FiPL Coordinator), supported by Heather Harley, the Farming and Land Management Officer, and Alex Farris, the Natural Environment Manager.

Charlotte, Sarah and Vickie all combine farming with being in the FiPL team.

Charlotte grew up in Exford and graduated from the University of Reading with a BSc (Hons) in Rural Resource Management. She was a volunteer with the ENPA before getting a job as an advisor for FWAG for 15 years. Since 2006, she has lived in Rodhuish with her husband Paul on a mixed farm with arable, beef and sheep. She joined the ENPA in 2017, working on several projects – the most recent being FiPL.

Below: The FiPL Team L-R: Vickie, Charlotte and Sarah. *(EHFN)*

"Working part time and job sharing has given me the flexibility to help out on our family farm and be with our three children while developing my own career,"

Charlotte says. "I really enjoy being in the FiPL team and having the opportunity to work on Exmoor again."

Sarah lives at West Ilkerton Farm, near Lynton, where she has a flock of her own sheep and helps her parents with their livestock.

She has always loved farming and the countryside. She studied Countryside Management at Bicton College before gaining a BSc (Hons) in Agriculture and Farm Business Management from the Royal Agricultural University, Cirencester.

Returning home to the farm, she did relief farming, helped out at home and worked part time in the National Park Information Centre at Lynmouth.

"I enjoyed working there and sharing my love of Exmoor," Sarah says. "But the job of FiPL officer came up and sounded really interesting, so I applied for it and was delighted to join the team. I love being part of the farming community, helping farmers and seeing projects come to fruition."

Vickie grew up on a tenanted sheep farm in the North West of England. She studied Animal Science with Livestock Production at the University of Wales, Aberystwyth, graduating with a first-class degree. After that she spent 12 months working on dairy farms and as part of a shearing gang in New Zealand before returning to manage an organic dairy herd.

Above: Charlotte. *(Paul Thomas)*
Below: Sarah. *(VE)*

She came to West Somerset in 2016, when her husband Andrew moved back to the family farm near Brompton Ralph, where they farm suckler cows and sheep. Prior to joining the ENPA, Vickie worked for the Quantock Hills National Landscape team in their Landscape Partnership Scheme and FiPL teams.

Since the FiPL scheme started, many Exmoor farmers have been able to carry out a variety of projects that have made their farms easier to manage and better for wildlife. The Nicholas family, for instance, restored an old hedgebank that had fallen into disrepair – conserving a historic landscape feature and creating some happy memories as well.

Left: Vickie. *(SE)*

Lesley and Paul Nicholas
Girt Down Farm, Combe Martin

Lesley Nicholas remembers exactly when she met her husband Paul. "We were fourteen and turkey-plucking for Grampy Booker. We married on Paul's twenty-first birthday, and we'll both be sixty-two this year."

They've achieved a huge amount in 40 years, raising their three children – Jake, Kingsley and Lizzie – while building up a farm business from scratch. They bought some land at Girt Down and lived in a caravan before building their own bungalow there in 1990.

They also bought Patchole Manor and rebuilt it, and have helped their children set up their own businesses. Lesley feels passionately about encouraging young people into farming, and says it was the main reason why she became a director of the EHFN.

"At home, I tend to have the bright ideas and Paul has to sort them out practically," she remarks, smiling. One of her bright ideas was to open a campsite during the 2020 lockdown, when there was a shortage of holiday accommodation. It's proved to be a great success.

"Paul loves talking about farming to the visitors," Lesley says. "He's the farmer, really; I've spent most of my life catering. He and Jake farm together now, but Jake's also got White Causey Farm, near Arlington."

They farm about 450 acres in all at Girt Down, with additional seasonal grazing on NT land by the sea. They keep up to 500 ewes (mostly Beulah Speckled Face) and roughly 50 cows (Limousin crosses and a few Devons).

The family are also keen on horses. They have bred some successful racehorses recently, which has been exciting for all concerned.

There's not much shed space, so they lamb the sheep outside in fields near the coast. "We've tried bringing them to the other side of the farm where it's more sheltered, but they like the fields they think of as home," Lesley explains. "They're very good mothers and tuck their lambs under a hedge in bad weather."

It was while watching Paul trying to catch ewes in a couple of fields with a broken-down hedge, through which the sheep always ran, that Lesley decided to apply for a FiPL grant to restore the stone-faced bank and plant trees on top.

"We didn't want to cut off the existing blackthorn trees, so it was a difficult job casting up the bank with fresh earth to stabilise them. We put new trees in the gaps and did the stone walling in February 2023," she says. "The weather was so good when we were making the wall that we had picnics outside. It was a lovely family occasion we'll remember every time we look at it."

Left: Paul and Lesley.
Below: Lesley by the new hedgebank they restored with the help of a FiPL grant. Combe Martin is in the background.

The EHFN Pre-Lambing Breakfast
Winsford

An event that's fast becoming a tradition is the Network's pre-lambing breakfast. This year it was held on Sunday 25th February in Winsford Village Hall, and the atmosphere was buzzing from the moment the doors opened.

Shirley Julian from Court Place Farm, Skilgate, was head chef. She has a catering business called Country Catering and also works in the café at Cutcombe Market. "My previous record was catering for ninety full English breakfasts, but this morning over a hundred and fifty people turned up!" she said as she was having a well-earned break afterwards with her helpers Judith Fewings from Dunster, Tracey Speed from Carhampton and Caroline Case from Withycombe. The ladies are all friends and keen members of the EHFN and its Women in Farming Group.

Many other people helped, too. Bridget and John Goscomb provided nine trays of eggs from Sindercombe Farm while Mole Valley Farmers gave a generous discount on sausages and bacon.

"It's wonderful that so many people came along," Katherine Williams said. "We organised it because we wanted to bring everyone together before the main lambing and calving season starts, and I think that was very much appreciated."

A grand total of £1,191 was raised for the FCN and Royal Agricultural Benevolent Institution (RABI) for their work supporting farming families.

The date has already been set for next year's breakfast.

Top: Shirley Julian cooking multiple breakfasts.
Above: Edna Hayes selling raffle tickets.
Left: Winsford Village Hall was packed throughout the morning.

The EHFN Next Generation Practical Lambing Workshop
with Tom and Leah Cole at South Heasley Farm, Heasley Mill

In mid-February, there was a lambing workshop for the Next Generation EHFN Group at South Heasley Farm, which is farmed by 33-year-old Tom and Leah Cole in partnership with Leah's parents Graham and Janet Jerrett.

They own 150 acres and rent a further 35, and have built up their flock of predominantly Mule ewes from 57 in 2018 to 400 in 2024. Lambing takes place in February, to make the most of higher prices for weaned lambs before the autumn.

Tom also works part time for a local farmer, and the Coles are very grateful to the families of Raymond and Dennis Jones, who farm in the same parish and have given them a huge amount of help and support.

The workshop was led by Rachel Forster from Market Veterinary Centre in South Molton, and it covered many different topics, including the importance of good preparation, how to deal with common problems, the care of ewes and newborn lambs and an update on Schmallenberg and bluetongue, which are viral diseases transmitted by biting midges. Both are becoming increasingly common in the UK, probably due to climate change and trade patterns.

Everyone left the evening with useful information and practical ideas that will help them to have a successful lambing season.

Above: Tom and Leah hosting the Next Generation lambing workshop at their farm.
Below: An impressive setup inside the lambing shed.

The Exmoor and District Deer Management Society

Every year in the middle of February, the Exmoor and District Deer Management Society conducts a deer count on two consecutive days to help with the management of Exmoor's wild deer.

Charles Harding and Charles Parker organised the first count in 1994, in response to a deer appraisal for the NT that had missed some areas but extrapolated numbers to reach a figure of around 3,000, which they reckoned was too high.

"We did counts in 26 areas back then. There was a good supervisor in each area who got knowledgeable volunteers to help, and we had about 270 people in all," Charles remembers. "The total count for that year was 2,376.

In 1998, Arlington was added to the count, followed by Grabbist in 2007 and Clatworthy Reservoir plus the Brendon Hill Area in 2015. Even allowing for this, there has been a massive rise in the number of deer to 3,700 in 2023.

Several things have happened in the past 30 years that seem to have had an impact on the distribution and numbers of deer on Exmoor. Charles explains that there was more for the deer to eat after the removal of the headage payments, which meant they bred at a younger age and moved into new places. Also, since the hunting ban and the right to roam, the deer seem to have moved from woodlands that were their stronghold, like Horner, to higher moorland areas like Dunkery – probably so they can see what's going on. However, they still go down into the valleys in bad weather.

Certain stags will travel miles, whereas hinds tend to be hefted to an area, moving within it according to the seasons and the weather.

"We need to keep deer numbers to a reasonable level, then people will tolerate them more," Charles says. "Recent TB research has highlighted the problem of deer and bovine TB – especially in the Dunkery area – and deer will also be a potential problem with all the new tree planting that's planned. As with other issues, like moorland grazing and swaling, a happy balance is needed."

Charles has always been fascinated by Exmoor's red deer. His mother was Master of the Devon and Somerset Staghounds, and his childhood was spent riding, harbouring and hunting.

After several years gamekeeping and managing wildlife, the NT employed him to manage their deer after staghunting was banned on their land in 1997. His work for the NT gradually increased, but last year he left and is now working for the ENPA catching grey squirrels.

Top: Charles and Duncan planning their route.
Left: Well-behaved horses are the best form of transport for counting deer on Exmoor, and binoculars are essential.

"Squirrels cause a huge amount of damage to trees," he explains. "You can fence against deer but not squirrels."

When Charles sent out his annual email to deer counters in at the beginning of January, he asked them to recruit young people. With this in mind, he asked Duncan Westcott to take part.

"Farming and hunting on Exmoor are interlinked, and I grew up with both," Duncan says. "I've always been interested in the red deer, so when I was asked to become involved with the count this year I was more than happy to help. Charles is the best possible person to learn from, and it's ideal seeing deer from a horse as you can pass through them without spooking them."

Duncan grew up on Exmoor and studied Agriculture at Bicton College and the Royal Agricultural University, Cirencester. After that he travelled to Australia and New Zealand for a year before coming back to Exmoor to do self-employed farm work – apart from five years in the construction industry.

"I learned a lot about the house-building process, and the wages were far better, but I was drawn back to farming, and that's what I'm doing now."

He is building up his own sheep farming business with Welsh Mountain and Welsh Mule ewes, based at Brockwell Farm near Wootton Courtenay, which is owned by Chris and Dee Binnie. In addition, he does other contracting work on local farms and fits in staghunting when he can.

Charles feels it's important that the deer count on Exmoor remains independent, despite pressure from the government and large organisations who want to become involved. "If it isn't broke, don't mend it," he says. "There's a huge amount of practical expertise here, and people give their time because they love doing it. The deer bring the Exmoor community together."

Above: Three young red deer. *(Shaun Davey)*
Below: It takes a lot of skill and experience to count deer accurately, even when they're out in the open.

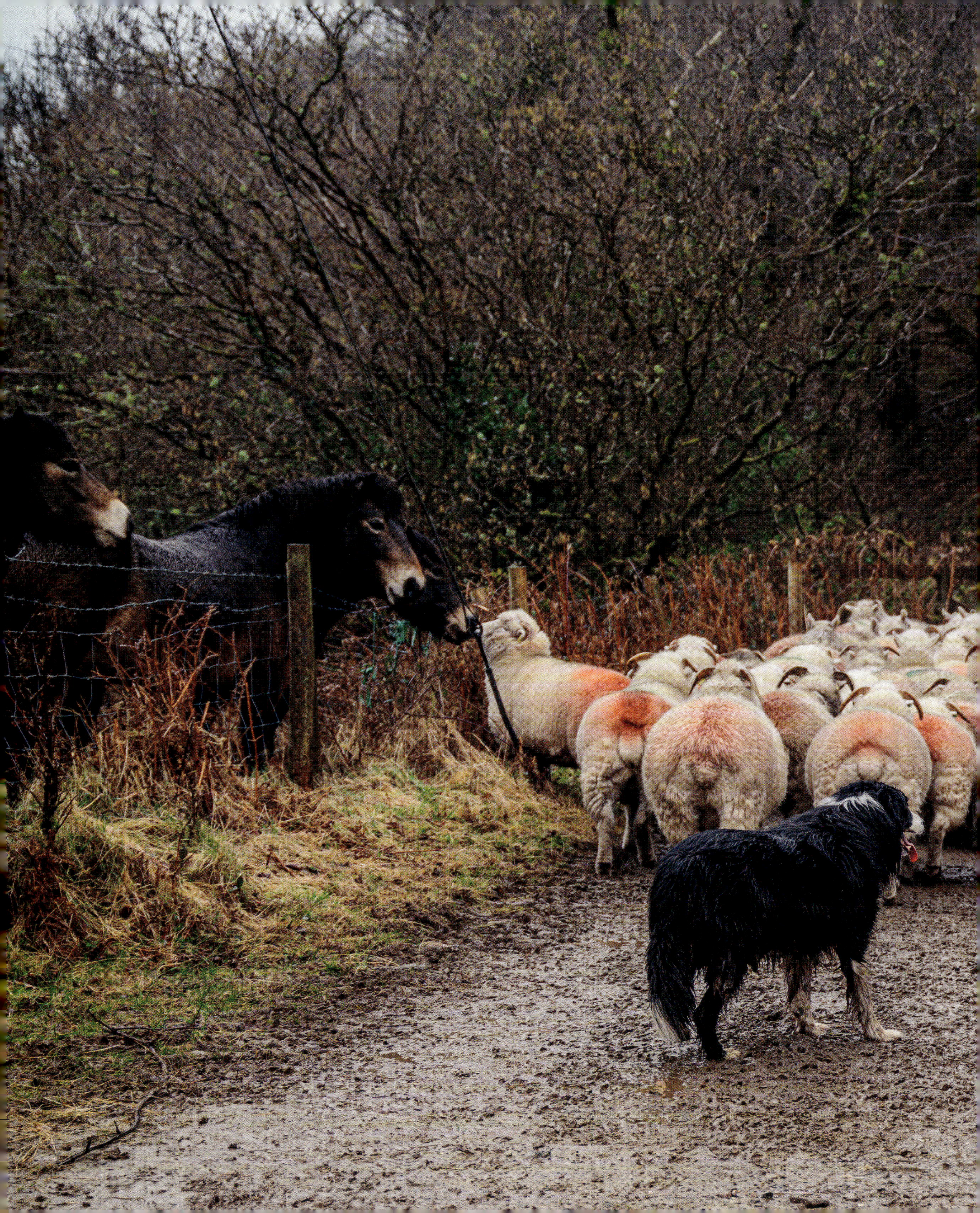

Sheep going back to the farmyard for scanning at West Ilkerton Farm.

March

It's said that March comes in like a lion and out like a lamb, but this year Exmoor had its most wintry weather at the end of March, with blizzard conditions for a while. The rhythm of life carried on regardless, though, as calving and lambing began in earnest.

At Great Champson, Richard and Kim Dart combine calving and lambing with getting some of their best pedigree Devon bulls ready for the sale ring.

John Prideaux also keeps Devon cattle; he thinks regenerative agriculture using quality Devons is a good way forward for farming on Exmoor.

Some areas, however, are capable of sustaining more intensive agriculture. The Speeds grow a variety of crops on their farm. Andrew Speed is passionate about the importance of food production as well as conservation, and he is one of the Exmoor farmers who are engaging with young entrants into agriculture and also policymakers at a regional and national level. Two others are Chris Webber and Robin Milton, who are representing farmers in a variety of ways, from local committees to national government, working for the future of farming.

Spring is just around the corner.

The Dart Family
Great Champson Farm, Molland

During the Napoleonic Wars, a lot of breeding livestock were sold for meat because food prices soared, but fortunately a few breeders of Devon cattle resisted the temptation to sell. One of them was Francis Quartly of Great Champson, Molland, who carried on improving his herd with the best cattle he could find.

The Quartly family continued to breed the highly prized 'Quartly' or 'North Devon' type of Devon cattle – suitable for farm work, milking and meat – throughout the nineteenth century. The working oxen at Champson were said to be the best in the country.

Today, the tradition of breeding Devon cattle at Great Champson is being upheld by the Dart family. George Dart signed a tenancy agreement in 1945, and his son William took it on in the 1980s. William has now taken a step back, and his son Richard and daughter-in-law Kim do most of the day-to-day farming.

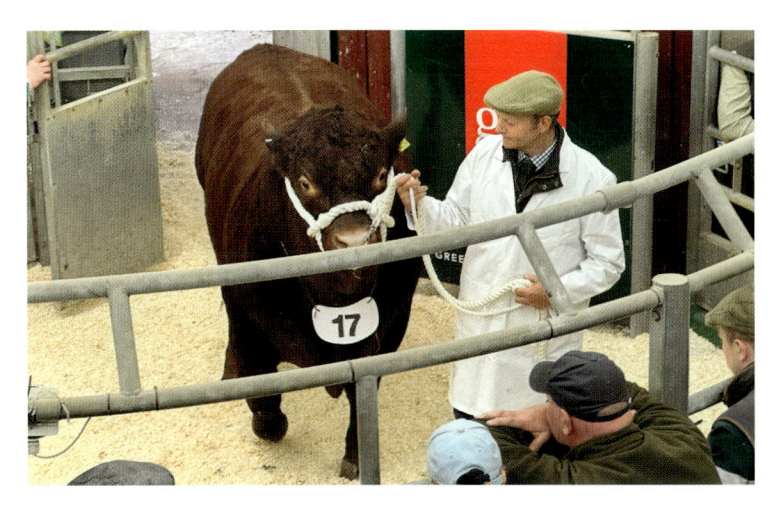

Above: Richard entering the sale ring with Champson Bullion, who was sold to Keith Francis for 5,700 gns. *(VE)*
Below: Some of the Dart family's Devon cattle with Great Champson in the background.

Top left: Kim bottle feeding a lamb.
Top right: A Closewool ewe and her twin lambs.
Above: Teamwork, Richard and Kim forking in silage for the cattle.
Left: Calves often jump through feed barriers and snuggle down in the silage or hay.

"We like the local breeds because they're well-suited to the conditions here, and they produce good meat," William comments.

"And when you've got cattle with a history like the Champson herd, there's a sense of responsibility to keep that going," Kim adds.

They keep about 40 pedigree cows, a bull and youngstock. About four bull calves from each year's crop are selected to be reared and sold as stock bulls.

"We train them to lead and keep them clean and separate from the rest of the herd in individual houses in the old farmyard. They're mucked out, fed and watered by hand, so it's a lot of hard work," Richard explains. "You've got to have depth of breeding if you're going to breed a bull – temperament as well as conformation."

"And if you want good livestock you've got to have good food," William emphasises.

This year, Richard took three bulls to the Devon Cattle Breeders' Society (DCBS) Spring Show and Sale at Sedgemoor Market. One of them, Champson Premium, won Male Reserve Champion, and all three bulls sold well.

The Darts also keep around 250 Devon Closewool ewes, 100 Exmoor Horn ewes and 100 Exmoor Mule ewes. Great Champson consists of about 300 acres of in-bye ground just outside the National Park and 100 acres of improved moorland within it.

"If we don't stock that high ground it fills with deer; our Exmoor Horns and Closewools do well up there," Richard says. "We breed a few Texels, Suffolks and Chartexes, too, for the rams, but you get to lambing and wonder why! The local breeds are so much easier."

Kim and Richard say they enjoy lambing and calving, even though it's hard work. "Sometimes we only get four hours sleep a night, but it's rewarding when the lambs and calves go on and do well," Kim remarks.

The family has taken an active interest in the three local breed societies over the years, and have held various posts, including Chair, Secretary, President and Member of the Council. In addition, Richard and William are bull inspectors for the Devon Cattle Breeders' Society.

Above: Kim with Fanny the Border Collie and Richard with Rosie the Terrier. *(Amanda Lockhart)*

"There's a lot of work involved, but you have to be pro-active," Kim says. "You can't just sit back and expect things to happen."

Richard and William agree, commenting that breed societies play a vital role in maintaining, promoting and marketing the breeds.

The Darts are cautiously optimistic about the future. "I think ELM is making us look at what we're doing in a new light. We've got to become more extensive whilst maintaining profitability," Kim says. "And maybe showing would be a good way to promote our stock." Livestock from Champson have had a good reputation for centuries, but in the competitive world of pedigree livestock breeding, success in the show ring is a tremendous advantage.

John and Pippa Prideaux
Croydon House Farm, Timberscombe

John and Pippa first saw Croydon House in April 2000, fell in love with it and bought it at auction the following week. "It was completely insane, but falling in love is completely insane," John remarks. It took 10 years hard work before they could move in properly.

Restoring the seventeenth-century house and its 15 historic farm buildings was a challenge.

"Our engineer told us the only thing holding the house up was force of habit. Of course, there were planning issues, but the National Park was very supportive."

Croydon came with about 330 acres when the Prideaux bought it. There are about 570 acres now as land sold away has been bought back. "We try to do right by the place. We respect the historic landscape in the same way as the house and buildings," John says. "Boundaries are the brushstrokes on the landscape, and there are about twelve kilometres of them. Some field names are over a thousand years old."

For a few years the Dascombes' cattle grazed the land, before the Prideaux took the farm in hand. There were Devon cattle and sheep initially, and Trevor Ball looked after the sheep.

John and Pippa started their pedigree Devon herd in 2006 with a few heifers and Whitefield Explorabull, who had won Young Bull of the Year. It has not been entirely easy. Half the herd were lost to TB from 2007-2010, followed by bovine viral diarrhoea (BVD) in 2011. The herd now has high health status. There are currently 130-180 cattle, looked after by Russell Liversidge and John.

"The cattle do well on grass, without any fertiliser or bought feed. Devons have evolved to do well on Exmoor's relatively poor grazing and steep ground. Regenerative agriculture using quality Devons is a good way forward for Exmoor farming," John comments.

A forest school (Wild Wellies) is also based on the farm.

John and Pippa are both from West Country families, but their previous experience of Exmoor and farming was limited to John spending family holidays in a cottage at Woolhanger. He particularly remembers helping George McCracken with sheep dipping and haymaking. They have learned farming on the hoof, and from the people they have worked with.

From an early age, John was fascinated by railways, the countryside and conservation. He wrote the history of the Lynton and Barnstaple Railway when he was still a teenager. He was Head of Policy at British Rail, Managing Director of InterCity and Chairman of Union Railways,

Below: Pippa and John beside one of the old traditional farm buildings.

Below: A freshly calved cow.

Above: Devon cows and calves in the spring sunshine.
Below: Croydon House Farm.

where he took the Channel Tunnel Rail Link to St Pancras and learnt a lot about balancing different environmental aims. He is Chairman of the Ffestiniog and Welsh Highland Railways in Snowdonia – part of the latest British World Heritage Site.

Pippa breeds competition dressage horses and spent much of her career coaching senior people to communicate effectively.

"We really admire the way in which the local farming community pulls together. They have helped us in all sorts of ways," John says. He was very pleased to be asked to be a Director of the EHFN and a member of the FiPL assessment panel, where his project appraisal and management skills are put to good use. Pippa was a governor of Cutcombe and Timberscombe schools for 10 years – Chair for part of that time – and is a Trustee of Home Start.

John thought Exmoor's Ambition was a triumph of cooperation between farmers and others on Exmoor. It was a blueprint which allowed all sizes of farm to develop sustainably. He is not a fan of some aspects of ELM, especially Landscape Recovery. "Only large, rich estates and NGOs are likely to benefit – who else can commit to twenty years or more?"

However, he stresses that we must make our farms work with what is on offer rather than waste time wishing things were different.

Andrew and Tracey Speed
Briddicott Farm, Carhampton

The Speeds rent about 2,500 acres on the Dunster Estate, which was purchased by Lord Hintze from the Crown Estate in 2017. Nearly all the land they farm lies within the National Park boundary, and it's unusual in that about half of it is suitable for crops.

March is a time for new life, with calves being born, spring crops like oats and beans being sown, and winter wheat, winter barley and oilseed rape all waking up after the winter.

"It's such a lovely area here – we're so lucky," Tracey says. "I especially like the new leaves appearing on the beech hedges at this time of year."

"We've been lucky," Andrew agrees. "But what we've done has been led by ambition and passion. I don't want to be in the position of looking back at my life and saying, 'I wish I'd had a go at that'. I rarely turn down an opportunity, but try to calculate out the risk. I'm useless at words – I think in numbers."

When he was 15, his father died and his family was left in turmoil. By the time Andrew had finished studying at Cannington College, his brothers were doing other things and he ended up farming with his mother on their 60-acre County Council farm in West Somerset. Milking sheep was almost unheard of back then, and Andrew thought it would be a good niche market to get into. Unfortunately, though, the cooperative he joined folded.

"We never saw a penny from all the milk we supplied, but I stuck with it and started supplying milk direct to processors – including David Baker for his Styles Ice Cream business," he remembers. "It was the lucky break I needed, and I ended up share farming with him. Sheep milking opened the doors to many other things."

Below: Andrew, Rob and Tracey Speed on the lawns below Dunster Castle, where they graze their sheep.

During those early years of farming, Andrew also worked in a local pub to make ends meet. It was there that he met Tracey, who was also from a farming family, and they married 30 years ago. They have three children: Kirsty (25, working in farm conservation in Sussex), Rachel (24, working for the National Farmers' Union (NFU) as a specialist in international trade) and Robert (23, working on the farm).

"Robert's super-keen," Andrew comments. "We're on the same wavelength with everything – well, practically everything!"

The farm is more or less split in two: one part for conservation and wildlife, the other for producing as much food as possible.

"There are roughly four hundred acres of lowland heath on Dunster Deer Park, which we're aiming to restore

Above: A young workforce: Callum Sexton, Daniel Smith, Matthew Williams, Jade Ely and Rob Speed.
Below: Andrew sowing spring beans, with views over the Brendon Hills towards Raleigh's Cross. *(SE)*

Left: Hereford cows and calves.

by grazing, refencing, replacing conifers with deciduous trees and removing a lot of silver birch and gorse – as well as topping to encourage heathland plants like heather," Andrew says. "Then there's the rest of the farm, including about a thousand acres that are farmed fairly intensively under a rotation of different crops like wheat, barley, oats, spring beans, maize, oilseed rape and lucerne – a lovely drought-resistant crop for silage – the sheep go mad for it. Another crop we're trying is sorghum, which loves heat. We can't deny that global warming is happening; we've got to look at the new opportunities it will bring."

He goes on to say that at last the government seems to be realising food security is important, and that if we're going to make space for nature while still being able to feed the nation, land used for food production will have to be farmed as efficiently as possible.

Until five years ago, they were milking up to 400 sheep, but then markets for the milk changed and new farming opportunities arose. They now have 1,500 Poll Dorset ewes that lamb three times over two years – in January, May and September – which means they have a supply of lambs throughout the year.

They also keep about 200 suckler cows (43 pedigree Devons and the rest Herefords). Over half of them calve now, in March, and around 85 in the autumn.

Three full-time and two part-time workers, all under 30, are employed by the Speeds – both girls and boys. They have an apprentice and a work-experience undergraduate from Harper Adams.

"We love starting work early in the morning – it's the best part of the day," Andrew says. "The staff usually arrive by seven o'clock. They work for two hours before breakfast at nine, when we all sit down together and talk about many things. It's a time for the transfer of information and planning the day ahead. Everyone takes it in turns to do the diary, which is a great way to learn about what's happening on the farm."

The Speeds are keen to share their love of farming with young people.

"One of the most enjoyable days of the year is when we show the reception class at Dunster First School around the farm," Tracey says. "They come back in year three as well, and it's lovely how enthusiastic they are."

Andrew also visits Minehead Middle School and Danesfield School to talk about careers in agriculture. "Some teachers seem to think farming is for pupils who aren't particularly academic, but good farmers have to be bright, responsible and eager to learn," he says. "That's what I look for in the students I employ. Previous experience isn't important – in fact, it's often easier to teach someone from scratch."

He is keen to share his love of farming with adults, too. Fourteen months ago, the family hosted a fact-finding visit from Natural England at Briddicott, and now Andrew has been asked to speak at a conference in Exeter.

Despite all the uncertainties, the Speeds feel positive about the future.

"That's my weakness: I'm always optimistic!" Andrew says with a smile. "For me, farming's a passion. It's the most important industry in the world, and I think we sometimes lose sight of that."

Right: Dunster Deer Park, where deer and Exmoor ponies are helping to enhance biodiversity.

Robin Milton
Higher Barton Farm, West Anstey

The Milton family's Exmoor ponies, managed by Rex Milton, are featured in October. In addition to farming, older brother Robin's career has taken a slightly different course – partly due to a sheep that ran into his leg and ruptured his Achilles tendon.

"I couldn't walk for some time, which was incredibly frustrating," Robin remembers. "I was already doing a bit for the NFU's Upland Farming Forum, and I happened to see the ENPA was looking for Secretary of State Appointees. I applied and, much to my surprise, was appointed. I probably wouldn't have been now, as the Defra selection process favours civil servants nowadays."

After being a Secretary of State Appointee, Robin carried on as a Parish Council Appointee for East Anstey and then a District Council Appointee for the Bishops Nympton ward of the North Devon Council. He was Chair of the ENPA committee for six years, and is now Deputy Chair.

"It's been interesting, frustrating and challenging in equal measure, but it's an opportunity to represent farming, which is often under-represented on such committees," he says. "I think one of the high points was *Exmoor's Ambition*, which was a collaborative piece of work, between the ENPA, EHFN, ES and others, that generated a lot of goodwill and has influenced farming policy hugely – much more than a lot of people realise. The Nature Recovery Plan a while later was a challenge. In many ways it was two years ahead of its time, but the main lesson to be drawn from the conservation side is you've got to talk to the people on the ground."

Robin chaired the UK National Parks Committee for a year, the National Hill and Upland Farming Forum for eight years and the Southwest Farming Forum for 15 years. He is a Trustee of the Foundation for Common Land and helps Defra with policy work, and he is also a board member of the Devon Local Nature Partnership, which is putting together the Devon Nature Recovery Strategy. "I'm the only farmer representative – almost everyone else is the employee of an NGO," he says. "Luckily, the current Chair is Professor Michael Winter, who's very good indeed."

Above: Robin with his grandson Percy.
Below: Networking. *(SE)*

In general, Robin says the knowledge and understanding of farming amongst many of those responsible for making decisions about the countryside is "frighteningly small".

"We've certainly had an interesting succession of Defra ministers," he comments. "It's easy to make promises if you know you'll be moved before you have to implement them. If the countryside were as important to the government as it makes out, why haven't we seen an elevation in the importance of Defra? And why hasn't agricultural support been index-linked? At least Rishi Sunak finally admitted that food is a public good at the Oxford Farming Conference this year."

Despite a certain degree of cynicism, Robin is also a realist and has a practical attitude towards the changes that keep happening to farming policies.

"Change is inevitable – you have to embrace it or you'll lose out," he says. "Farmers have got to be prepared to work creatively, without the hang-ups of past policies. They must capitalise on every opportunity while keeping the base structure of their farms in place so that when things change again (as they will, sooner or later) they can adapt accordingly. That's really the definition of sustainability: keeping the business going year after year."

He admits that smaller farms are the most vulnerable. "The social structure of an area collapses when small farms go, and sadly that's a real challenge around Exmoor. I don't think it's all doom and gloom, though. Farming's a bit like life: full of ups and downs. But in the end it's what you make of it."

Apart from a runaway sheep, the factor that Robin believes was crucial in allowing him time to branch out into policy work was an early succession plan. "My son Buster became my boss when he was twenty-three, and that freed me up to do other things," he says.

Despite all the influential posts he has had, Robin is still a farmer at heart. Being a Director of the EHFN is very important to him, and what he likes most of all is being at home doing the farming.

Below: Robin giving a presentation at a Defra FiPL workshop, with South West FiPL teams, at Horner Farm. *(SE)*

Moles Chamber looking towards Muxworthy Farm. *(Debbie Tucker)*

Chris Webber
Hindon Farm, Minehead

Like Robin Milton, Chris had an injury that changed the course of his life. "I spent my teenage years in Porlock, and all I ever wanted to be was a Royal Marine," he says. "So I left school at sixteen and joined up, but a few years later I tore my knee cartilage and had to quit."

He trained as a lifeguard and worked at Aquasplash in Minehead, where he met Emily Webber from Hindon Farm. Soon he was helping out on the farm, and he loved it.

Chris and Emily were married in Selworthy Church by his mother, the Reverend Ann Gibbs, in 2021, and since then he has been farming in partnership with Emily and her parents Roger and Penny Webber. Emily has a job off-farm as a human resources manager, so Chris is responsible for the farm's day-to-day management. He took Emily's surname when they married.

Hindon Farm has approximately 500 acres of in-bye land, rented from the NT, plus 1,500 acres of coastal heath on

Right: Chris at Hindon Farm, with Minehead in the background.
Below: Aberdeen Angus cattle, with Grabbist Hill in the distance.

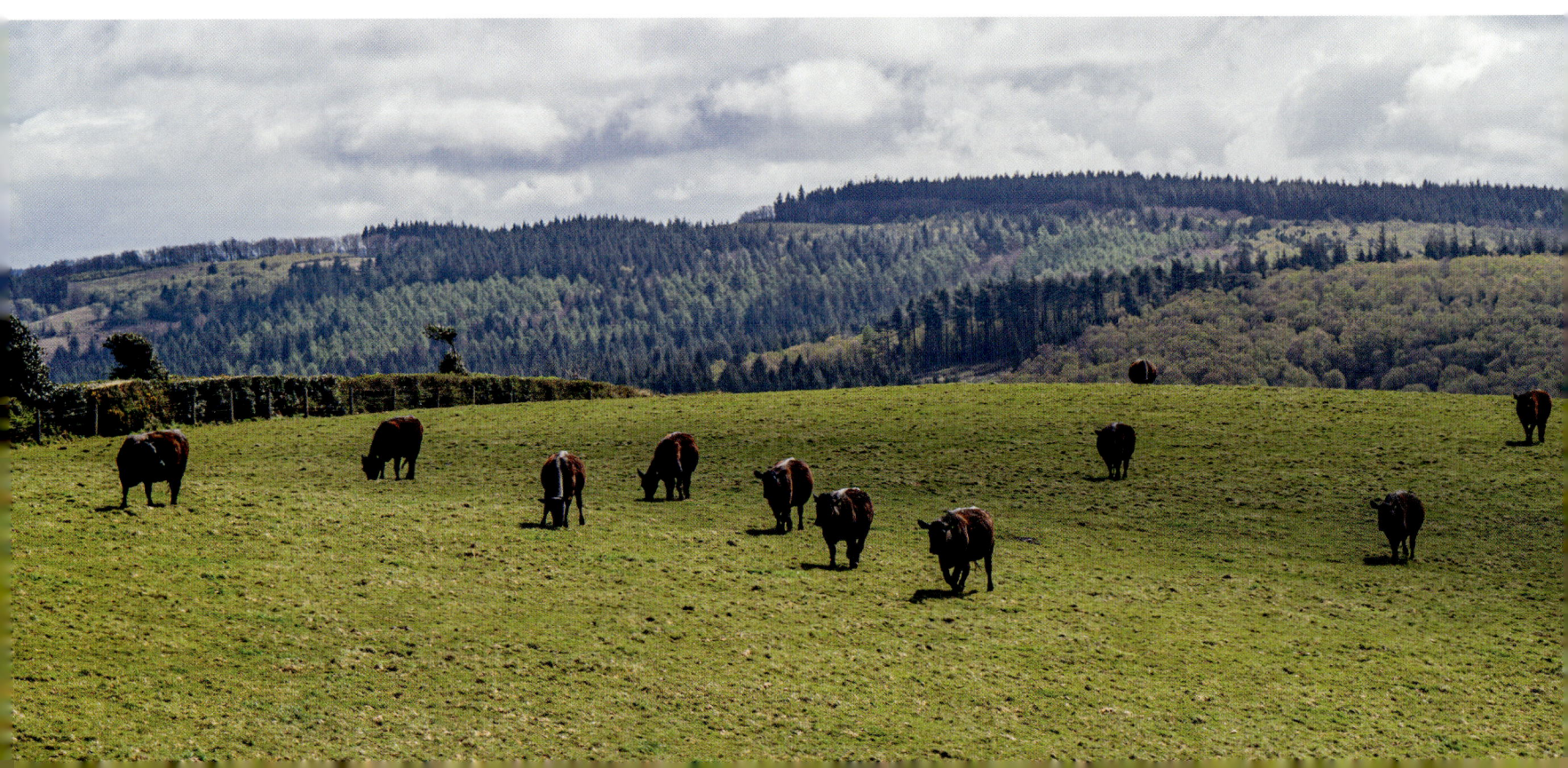

North Hill under a grazing licence: some owned by the NT and the rest by the ENPA.

The farm has been organic for about 24 years, and it is also certified with Pasture for Life.

The NT's Landscape Recovery Scheme is mainly taking place in Porlock Vale, but the Webbers are trying to get funding for surveys.

The farm has 150 New Zealand Romney and Welsh Mountain ewes that lamb outdoors to North Country Cheviot rams, and 50 Aberdeen Angus plus 50 Belted Galloway suckler cows. "The farm needs sheep and cattle, but it's the cattle I really love," says Chris. "I'm reducing the Angus herd and increasing the Belties. They have good temperaments, can live outside all year round and I've never had to help them calve. They take a little longer to finish, but their kill-out percentage is incredible: up to sixty per cent."

Chris has taken part in Allan Savory's holistic farm management course, and finds it's something he draws on to guide his decisions.

"Farming lost its way a bit and became a conveyor belt, but it's so rewarding to see the improvements that can be made in both the livestock and the land if balance is restored," he remarks. "Diversity in the soil leads to a healthy system with healthy livestock."

In the past couple of years, Chris has planted a hedgerow that was funded by a FiPL grant, and he's in the process of planting 10,000 trees.

A major change he has introduced is a paddock grazing system. At the moment up to 100 acres of hay or silage are made per year, and the cattle are grazed in 0.5 – 0.75 acre paddocks and moved twice daily. Each paddock gets about 80 days of rest between being grazed, but the aim is to have two grazings per year and not feed any forage.

Away from the farm, Chris has been in the Minehead Lifeboat crew for six years and is a trainee helm. He has also become involved with the NFU – another thing he and Robin Milton have in common.

"I went to a couple of meetings. Dave Knight was Chair of the West Somerset Group, and before long I was asked to be Vice-Chair. Dave stepped down when he became too busy with his young family, and now I'm Chair."

He is also on the Tenants' Forum for the South Region NFU and is the Somerset Representative for the Uplands Group.

"Farmers are a very diverse group of people, but as an industry we have to try to speak with one voice," Chris says. "In the past we've had subsidies for producing food, with environmental subsidies on the side. Now the BPS (which was essentially a food subsidy) is ending, and the squeeze is going to be on farming as food prices rise. Farming is a lifestyle – it absolutely is – but it's also a job. We need a fair price for producing food. Everything else is an extra."

Below: A Belted Galloway cow on Bossington Hill. *(Shaun Davey)*

The Reverend Prebendary David Weir
The Rectory, Exford

David grew up near Chichester in West Sussex, where he spent a lot of time sailing. He remembers being attracted to a religious life from an early age, thanks to a vicar who used to visit his primary school.

"I felt very drawn to the atmosphere in assembly, and I loved the thoughtful, reflective, creative space the Church provided," he says. "We were never invited to believe in a God that intervenes, but a God that is ever-present. That's so important."

His first jobs for the Church of England were in the Portsmouth Diocese, after which he took a break, went to Cornwall and studied for a degree in English Literature.

"It reinvigorated my interest in Biblical texts. I love reading, studying and absorbing new ideas. The idea that words are interpreted by different people in different ways struck a particular chord in me and continues to help me in what I do now."

An advertisement in *The Church Times* for a part-time, four-year post as Priest-In-Charge of the parishes of Exmoor, Exford, Hawkridge and Withypool brought David to Exmoor National Park in 2009.

"All the members of the four Parochial Church Councils were invited to Simonsbath Church to meet the candidates the day before the interview, and I immediately felt a tremendous sense of community and friendship," he recalls. "The interview was charming – very enjoyable. It was held in Edna Clatworthy's dining-room, with two gigantic sets of antlers on the wall. There were eight Churchwardens (Edna, Beryl Scoins, David Bawden, George Vellacott, Kathryn Tucker, David Newman, Brian Duke and Anne Chown), the Rural Dean, the Lay Chair of the Deanery Steven Pugsley, and the Archdeacon, but it wasn't at all scary. I remember Brian smiling at me encouragingly all the way through. When I was offered the job, I got the distinct impression it was an invitation to listen, learn and join in, and that's what I've tried to do."

In the spirit of joining in, David accepted an offer from Leo Martin of riding lessons on her horse.

Top: David riding Sundae. *(Leone Martin)*

Left: David has been the Assistant Steward for the Exmoor Horn classes at Dunster Show for over 10 years. Mike Rawle (with his back to the photographer) is the Steward.

"How appropriate that the horse is called Sunday," a friend remarked.

"Even more appropriate that it's spelt like the pudding!" David replied. A fondness for cakes and tea is, he admits, a great advantage if you're a rector.

With Sundae and Leo's help, he reached a level of competence that allowed him to discover Exmoor on horseback, but his spare time became limited in 2011 when Winsford, Cutcombe, Exton and Luxborough were added to his original four parishes and he became Rector of the Exmoor Benefice.

In all, the Benefice covers around 120 square miles and has a population of about 2,000 adults.

"There are a few people spread over a large area, but I get to know my parishioners in a wonderful way that urban priests can't." David comments. "People on Exmoor are capable and independent, and they're used to travelling between parishes for all sorts of get-togethers, including Church services. Gathering with others is vital, especially in times of crisis or celebration. Singing in the Ring, for example, was a glorious celebration. Katherine and Charmain are so good at listening and providing what people really want."

Besides religious services, David can be seen helping at many other events – including Dunster Show, where he has been the Ring Steward for the Exmoor Horn sheep classes for several years.

His involvement with Exmoor's farming community has refreshed his appreciation of a favourite quote from *Howards End* by E M Forster: *In these English farms,*

Above: Sarah Atkins and David peeling carrots for the meal after Singing in the Ring. "When I was training for the ministry, a parishioner said, 'Just because you're ordained, you don't exchange your hands for wings.' I've never forgotten that," he says. *(EHFN)*

Below: The snowdrops that herald spring.

if anywhere, one might see life steadily and see it whole, group in one vision its transitoriness and its eternal youth, connect – connect without bitterness until all men are brothers.

David also treasures Hope Bourne's books and, as she did, loves the snowdrops that herald spring. "In Church, we gather round God in faith," he says. "Central to that is the resurrection – the mystery of new life when things seem at their bleakest. That is what happens every spring: new life and new hope after the bleakness of winter. This is a truly wonderful time of year."

The Team

Katherine Williams
Project Manager

Katherine grew up at Lower Blackland Farm with her younger brother Mark.

Her first job was with Stags at South Molton Market, followed by three years with Trading Standards. After that, she worked in estate agency before getting a two-year contract as Network Officer of the Exmoor Hill Farming Network.

Now in her tenth year as Network Manager, Katherine has been instrumental in the success of this inclusive, caring farmer-led organisation that has made a huge difference to so many people. She has also delivered programmes on behalf of The Royal Countryside Fund and has represented the EHFN at national level on several occasions.

When she isn't working, Katherine enjoys walking with her Jack Russell Terrier, Bonnie. She loves Exmoor and its wildlife, especially the deer, and her favourite place is the Barle Valley.

Victoria Eveleigh
Author

Victoria ("Tortie") and Chris Eveleigh have been farming at West Ilkerton, near Lynton, since 1986. They breed pedigree Devon cattle, Exmoor Horn sheep and Exmoor ponies, and have a couple of small-scale renewable energy projects as well. The farm is open for tractor-drawn farm tours during the summer.

Their son George lives in Scotland with his family, where he manages a sporting estate. Sarah, their daughter, combines helping on the farm with working for the ENPA as a FiPL Officer and a Project Manager for the Reviving Exmoor's Heartlands project.

Victoria has written nine books, illustrated by Chris, that have been published by Orion. She is also a regular contributor to the magazine *Exmoor*.

Chris is a Director of the EHFN.

This view from their farm includes a favourite landmark called Saddle Gate.

Eleanor Davis
Photographer

Eleanor grew up on her family's farm in Chittlehampton, North Devon, and has a particular love for sheep. She likes tending to her family's flock of Mules, Herdwicks and one special Exmoor Horn called Winnie.

Eleanor has always enjoyed photography. From a young age she could often be seen with a camera in her hand, photographing everything from moments on the farm to her brother's football team.

She started Eleanor Davis Photography eight years ago, specialising in weddings and family photos. This business is run alongside full-time employment in the public sector.

Her work has been featured in the Farmers Guardian Heart of the Mart campaign and the South West Farmer.

She loves spending time walking on Exmoor and taking photographs of its captivating landscape.

Jane Pearn
Editor

Jane was brought up in North Devon; she has been connected with farming all her life.

On leaving school, she trained as a teacher and worked in education until 2019, teaching a wide range of ages. For the last fifteen years of her career, she taught Maths and English to adults.

Her three children – Amy, Duncan and Caroline – are all involved in farming. Duncan features in this book, helping Charles Harding with the deer count in February.

Jane has been a Director of the EHFN since its inception, and she also sits on the Exmoor FiPL panel.

She and her husband John live on a smallholding near Yeo Mill.

Her organisational skills and knowledge of Exmoor have proved invaluable in the creation of this book.

Acknowledgements

My heartfelt thanks go to all who have helped bring this book to fruition from the initial idea to the finished article.

To the Exmoor National Park Authority Partnership Fund, The Exmoor Society, the sponsors listed opposite, and others who have all made financial contributions.

To Victoria Eveleigh and Eleanor Davis who have shown dedication and professionalism throughout to produce beautiful writing and photography, whatever the weather or set back. They have continually exceeded any expectation.

Jane and I are especially grateful to Steven Pugsley and Naomi Cudmore, for their help and encouragement. Thanks also to the additional photographic contributors: Emily Fleur, Sarah Hailstone, Shaun Davey, Debbie Tucker, the Eveleigh family and numerous others, and to Jen Brookes for her drawings which appear at the start of every chapter.

I am indebted to Exmoor Farmers Livestock Auctions and White Rock Cottage for allowing the launch and exhibition venues, and to the volunteers for assisting.

To Mark Couch, Kevin Bovey and the rest of the team at Short Run Press for their professional services to design and print the book.

I would also like to thank the farmers, landowners and individuals who have allowed us onto their farms, and into their homes and businesses, with such warmth and generosity. Without their time, this book would not have been possible.

Finally, thanks go to Jane Pearn who has gone above and beyond to voluntarily oversee this record and celebration of true Exmoor farming life. We will be forever grateful for everyone's help in capturing these moments in time which we hope will be cherished for many years to come.

Katherine Williams,
Exmoor Hill Farming Network Manager

Highland cow above Chetsford Water.

Sponsors and Grants

We would like to express our sincerest thanks to the following organisations, businesses and individuals for their donations. Without these, and other, generous contributions this book would have not been possible.

The Molland Estate, Exmoor

The Badgworthy Trust for the Preservation of Exmoor

Mr and Mrs David Wallace

Hollam Estate

Peter Carew

Robert Miles

May Brothers Ltd.

Woolhanger Farming Partnership

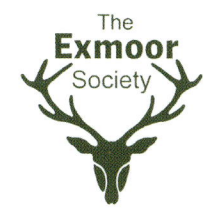

Glossary

Arable – the production of crops such as wheat and barley.

Cattle breeding – the process of developing a particular breed/herd of cattle by selecting for specific characteristics.

Continental breed – breeds of cattle (e.g. Charolais, Limousine or Simmental) or sheep (e.g. Texel or Zwartbles) that originated on continental Europe rather than in the British Isles.

Contractors/Contracting – the suppliers and use of external labour and machinery, usually for specific tasks such as tilling or harvest. For examples, see pages 41, 47 and 135.

Couple – a ewe and her lambs as a single unit.

Creep feed – supplementary feed fed to lambs usually before weaning.

Cull ewe – a ewe which has been selected not to be kept for breeding for another year.

Cutters and combs – the removable blades used in a shearing handpiece.

Dairy farming – the management of livestock, mainly cattle, but also sheep and goats, for the production of milk. See page 178.

Deer farming – the farming of domesticated deer for meat and other products. See page 58.

Direct selling – the sale of produce (meat and vegetables) directly from the farmer to the consumer, often through farmers markets and box schemes. For examples, see pages 58, 61, 143, 151, 154, 156, 158 and 160.

Exmoor pony gathering – rounding up ponies that have been pre-living on moorland so that the foals can be inspected, weaned and registered. See pages 121 and 126.

Farm school – the provision of agricultural and nature-based education for pupils and students through farm visits. See pages 69, 141 and 208.

FiPL – the farming in protected landscapes grants scheme funded by Defra and administered by the National Park. See pages 69, 141, 169, 192 and 194.

Fish farming – domestic fish production, often trout, in freshwater ponds. See page 154.

Fly strike – an infestation of blowfly maggots on a sheep.

Free-range poultry – Domestic fowl that are farmed for their meat or eggs with access to outdoor space. For examples, see pages 106, 110.

Gimmer – an unmated female sheep under 2 years old.

GPS electronic collars – collars used to manage livestock in unfenced areas often moorland. See pages 34 and 169.

Grazing management systems – principles and technologies used to make the most appropriate use of grassland. The system used depends on what the farmer is trying to achieve.

Handpiece – hand-held clippers used to shear sheep.

Haymaking – the production of dried grass that can be stored for feed in winter without rotting. For example, see page 41.

Headage payments – historic farm support payments paid based on the number of cattle or sheep being farmed.

Hefted/Hefting – the learned behaviour of sheep to remain in, and return to, an area of open grazing, often a moor or common. Often inherited from mother to daughter over multiple generations.

Hedgelaying – the traditional management of hedgerows by partially cutting upright stems (steepers) and laying them flat on the bank/ground, providing an effective stock-proof barrier and encouraging regrowth. For example, see page 135.

Herbal ley – an area of sown pasture made up of a mix of grasses, herbs and legumes.

Livestock judging – the judging of livestock in competition to identify the best examples of their breed and best quality. See page 94.

Markets – locations where livestock are bought and sold. See pages 23, 30, 77, 101, 115 and 117.

Min tillage – the cultivation of land without ploughing, designed to minimise disturbance of the soil.

Mires restoration – the rewetting of areas of upland peat bog to increase water quality and ecological function. See page 135.

Moorland grazing and management – the maintenance of moorland habitat through managed grazing other practices such as swaling (controlled burning).

Mule – a crossbred ewe whose parents were both pedigree breeds. Often a cross between a hardy upland ewe such as a Scottish blackface or Swaledale and a more prolific Blue-faced Leicester or Border Leicester ram.

Native breed - breeds of cattle (e.g. Red Devon, Hereford or Galloway) or sheep (e.g. Exmoor Horn or Suffolk) that originated in the British Isles.

Next generation – events and activities aimed at younger members of the community.

Organic farming – farming following a series of principles as outlined by accreditation bodies such as the Soil Association. This includes strict rules on the use of artificial fertilisers and pesticides. For examples, see pages 34, 38, 61, 139 and 214.

Plastic recycling – the recycling of farm waste plastic largely used in silage production. See page 37.

Polled – The genetic trait of certain breeds or individual cattle and sheep to not grow horns.

Raddle – Oil-based paint applied to the belly of a ram at tupping time to enable the farmer to see which ewes have been served.

Rare breeds – individual breeds of livestock that are at threat of extinction as assessed by the Rare Breeds Survival Trust.

Rotational grazing/cell/paddock grazing – a grassland management technique whereby groups of cattle or sheep are only allowed to graze a particular field or parcel of land for a restricted period of time, allowing for improved grassland utilisation.

Shearing – the removal of wool from a sheep using hand or electric clippers. For examples, see pages 41, 49, 51, and 89.

Sheep breeding – the process of developing a particular breed/flock of sheep by selecting for specific characteristics.

Sheep dipping – the plunge dipping of sheep to manage external parasites. See page 119.

Sheep dogs – breeds of dog such as the Border Collie and Kelpie bred to aid sheep management. For examples, see pages 129 and 131.

Sheep scanning – the use of ultrasound to predict the number of lambs a ewe is carrying before birth. See page 181.

Sponging – The process of using a hormone-impregnated device to synchronise the insemination of ewes to either bring forward or condense their lambing period.

Standing grass – Grass that has been left to grow for the production of silage, haylage or hay.

Steer – a castrated male calf.

Stirk – a young bull, steer or heifer usually between 1 and 2 years old.

Stone walling – the traditional maintenance of wall and banks through the use of interlocking stones without cement. For example, see page 101.

Store cattle – cattle that are sold to continue growing elsewhere rather than for slaughter.

Suckled calf – a calf that is sold usually in the late autumn soon after weaning.

Suckler cow/herd – cows that are kept for the production of calves bred for meat production.

Terminal sire – a ram or bull from a breed that has been developed for its meat production properties.

Two tooth – a ewe in its second year with 2 adult teeth.

Topping – the removal of seed heads and weeds from grassland, usually in the summer using a flail mower.

TB testing – the regular testing of cattle to assess whether they have bovine tuberculosis. See page 79.

Tupping – the time of year when the rams are put with the ewes for mating.

Wether – a castrated male lamb.

Wild-flower regeneration – the use of seed collection and over sowing to regenerate wild-flower meadows. For examples, see page 66 and 68.

Wool-shedding – sheep breeds, such as Exlanas and Easycares, that have been bred to shed their wool without the need for shearing. See pages 19 and 56.

Winter v spring crops – the difference between when a crop is sown, either in the autumn and allowed to overwinter or in the spring.

Zero waste – a farming principle whereby all outputs and by products are utilised. See page 38.

Acronyms

BPS	Basic Payment Scheme
BSE	Bovine Spongiform Encephalopathy
CS	Countryside Stewardship
Defra	Department for Environment, Food and Rural Affairs
EFLA	Exmoor Farmers Livestock Auctions
EHFN	Exmoor Hill Farming Network, often referred to as the Network
ELM	Environmental Land Management
ENPA	Exmoor National Park Authority
ES	Environmental Stewardship
ESA	Environmentally Sensitive Area
ESCRA	Exmoor Farmers Suckled Calf Rearers' Association
FCN	Farming Community Network
FiPL	Farming in Protected Landscapes
FWAG (SW)	Farming and Wildlife Advisory Group (South West)
HLS	Higher Level Stewardship
NE	Natural England
NFU	National Farmers' Union
NGO	Non-Governmental Organisation
NT	National Trust
RABI	Royal Agricultural Benevolent Institution
SFI	Sustainable Farming Incentive
SSSI	Site of Special Scientific Interest
TB	(Bovine) Tuberculosis
YFC	Young Farmers' Club, often referred to as Young Farmers